Brooklands Books

☆ A BROOKLANDS ☆
'ROAD TEST' LIMITED EDITION

MG
MGF

Compiled by
R.M.Clarke

ISBN 1 85520 4797

Brooklands Books BROOKLANDS BOOKS LTD.
P.O. BOX 146, COBHAM,
SURREY, KT11 1LG. UK

A-MGFX1

Printed in Hong Kong

ACKNOWLEDGEMENTS

For more than 35 years, Brooklands Books have been publishing compilations of road tests and other articles from the English speaking world's leading motoring magazines. We have already published more than 600 titles, and in these we have made available to motoring enthusiasts some 20,000 stories which would otherwise have become hard to find. For the most part, our books focus on a single model, and as such they have become an invaluable source of information. As Bill Boddy of *Motor Sport* was kind enough to write when reviewing one of our Gold Portfolio volumes, the Brooklands catalogue "must now constitute the most complete historical source of reference available, at least of the more recent makes and models."

Even so, we are constantly being asked to publish new titles on cars which have a narrower appeal than those we have already covered in our main series. The economics of book production make it impossible to cover these subjects in our main series, but Limited Edition volumes like this one give us a way to tackle these less popular but no less worthy subjects. This additional range of books is matched by a Limited Edition - Extra series, which contains volumes with further material to supplement existing titles in our Road Test and Gold Portfolio ranges.

Both the Limited Edition and Limited Edition - Extra series maintain the same high standards of presentation and reproduction set by our established ranges. However, each volume is printed in smaller quantities - which is perhaps the best reason we can think of why you should buy this book now. We would also like to remind readers that we are always open to suggestions for new titles; perhaps your club or interest group would like us to consider a book on your particular subject?

Finally, we are more than pleased to acknowledge that Brooklands Books rely on the help and co-operation of those who publish the magazines where the articles in our books originally appeared. For this present volume, we gratefully acknowledge the continued support of the publishers of *Autocar, Automobile Magizine, Car and Driver, Car Magazine, Car South Africa, Complete Car, Performance Car, Sports Car International, Top Gear* and *What Car?* for allowing us to include their valuable and informative copyright stories.

R.M. Clarke.

Brooklands Books

CONTENTS

5	MGF is Go!	*Autocar*	Mar	8	1995
14	MaGic! Road Test	*Performance Car*	Nov		1995
21	F for Fabulous	*Sports Car International*	May		1995
24	MGF 1.8i Road Test	*Autocar*	Sept	20	1995
30	MGF	*Car South Africa*	Nov		1995
34	MGF 1.8i Road Test	*Automobile Magazine*	Dec		1995
37	MGF	*Car and Driver*	Mar		1996
38	Brothers in Arms - MGF 1.8i vs. BMW Z3 Comparison Test	*Complete Car*	Apr		1996
44	Clan Gathering	*Top Gear*	Apr		1996
50	Champs - MGF 1.8i VVC vs. Mazda MX-5 vs. Lotus Elise Comparison Test	*What Car?*	Sept		1996
56	A Great Sports Car Illustrated Long Term Test	*Car Magazine*	Apr		1997
61	MG Force	*Performance Car*	Jan		1997
66	Fun Boy Three - MGF 1.8i VVC vs. Mazda MX-5 vs. BMW Z3 Comparison Test	*What Car?*	May		1997
72	Purple Pros, Purple Cons - 1.8i VVC Long Term Test	*Autocar*	July	9	1997
75	Three-Tier Tumble - MGF 1.8i VVC vs. Lotus Elise vs. Caterham Seven Superlight Comparison Test	*Complete Car*	July		1997
80	MG FF - MGF Cheetah	*Autocar*	Mar	4	1998
84	Wind Cheetah	*Performance Car*	June		1998
88	MGF Buying Used	*Autocar*	May	8	1998

NEW MG

MGF

Enthusiasts have been holding their breath to see what the new MG would be made of. Now we know — and in the next 10 pages we chart the genesis of Britain's most important sports car since the E-type

Words by Steve Cropley

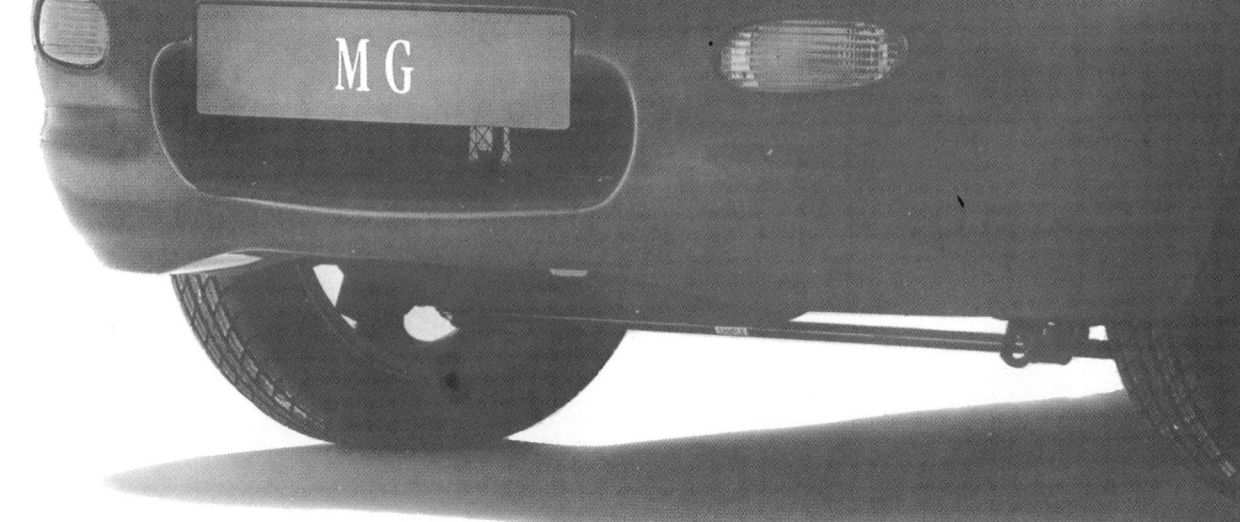

is go!

Photography by Stan Papior

NEW MG

Feast your eyes, car enthusiasts, on the sleek lines and rich specification of the new MGF roadster, revealed at the Geneva motor show today, and be glad that these are not the 'good old days'.

Why? Because the arthritic group of companies from which today's Rover Group has so remarkably sprung would never have built a car like the MGF. Because in the UK's new would-be sporting flagship Rover has given us practically everything on our MG wish list, whereas in the old days life wasn't quite like that.

On paper, the MGF is exactly the mid-engined, affordably priced, all-independent, all-disc roadster we have hoped for and speculated upon for years. Not only that, but it is also better packaged and more sophisticated mechanically than its near-rivals, the Mazda MX-5, Fiat Barchetta and Toyota MR2. Its styling studiously avoids anything of a 'retro' look, which might have pleased a few near-sighted purists but which made the MG RV8 such a disappointment.

The F-type introduces two new 1.8-litre versions of Rover's much-acclaimed K-series family of modular engines. The higher-powered version has an all-new variable valve timing system that gives it a horsepower per litre figure that would do justice to a modern turbo four.

And — amazingly — it brings Alex Moulton's time-tested Hydragas suspension system into the new generation of

Mid-engined MGF will start at £16,000 when it goes on sale in June. It promises much dynamically

Design

To Canley via Frankfurt

Rover's EX-E sowed the seeds of the MGF, says Iain Robertson

It's 10 years since Rover stormed a European motor show with an all-new MG sports car, but the success of the exotic EX-E concept unveiled at 1985's Frankfurt motor show was a turning point for the marque.

The EX-E was never intended for production – although some say it inspired Honda's NSX – but the essential MGF design cues were already there. There are mechanical similarities, too: the engine was mounted behind a two-seat cockpit and even the running gear, like much of the MGF, was borrowed from the Metro (the EX-E's 250bhp V6 started life in the raucous Metro 6R4 rally car).

By 1989 the MG programme had been dubbed "Phoenix Revival" and PR codenames assigned to a series of prototypes.

Spurred on by the success of Toyota's MR2 and the imminent arrival of Mazda's MX-5, Rover's special products department opted for the mid engine/rear drive layout of PR3, as the car was initially known. Dynamically superior to front-drive prototypes, it was also to be cheaper to build than traditional MGs with their front engine/rear drive layout.

Aware that its new mid-engined roadster would take at least four years to develop, Rover also sanctioned PR4 – better known as the low-volume MGB-based RV8 – to pave the way by re-establishing the MG marque.

In January 1991, with its own designers working flat out on models like the 600, Rover commissioned styling proposals from IAD (now part of the Mayflower Group that produces MGF bodies for Rover), MGA Developments and the Luton-based consultancy ADC.

Of the three, Coventry's MGA gave the MGF its basic shape. Former MGA designer Steve Harper recalls the brief. "It was very open, really. We were given a simple mechanical package and a list of cars that

1985's EX-E provided inspiration for MG rebirth

MGA's final proposal takes shape. Note air intake

'Twin cockpit' cabin enhances driver appeal and safety; power comes from 118 or 143bhp K-series 1.8

captured some of the spirit they were after – including the EX-E, the XJR-15, the original MR2, the Elan and BMW's Z1. Plus it had to be obviously British and unmistakably MG."

A delegation from Rover's in-house studio visited twice to follow MGA's progress – once early on to view initial sketches and again to see the full-size clay model.

Harper remembers their reaction. "The feedback we got was very positive. The high rear deck went down well, as did the bodyside surfacing and the car's squat stance, but the front end treatment drew some criticism – too anonymous, not MG."

Harper and his team were dispatched to nearby Styling International, where designers and modellers were putting the finishing touches to the RV8's familiar nose.

"We tried a similar look, with the round indicators and sidelights and the distinctive MG grille shape above the bumper line, but it wasn't fully resolved."

By the end of May – after just four and a half months – MGA had signed off its MGF proposal and the clay model was shipped to Rover's Canley design studio, where in-house designer Gerry McGovern started work on the production version. He insists that the final version is very different from MGA's 1991 proposal and he's right. The windscreen is new, the waistline has been lowered, every body surface has been altered and it's shorter and prettier. But the basic profile and many of the details are still there.

The most significant advances occurred at the front. By reducing the front overhang and raising the height of both front wings and bonnet, Rover's team has not only made space for a vertically mounted full-size spare wheel but created the classic MG 'face' that was missing from the MGA version. The rear is sharper, too – more like the EX-E – and every surface has been honed, reducing visual bulk and increasing the feeling of tension.

Effort to mimic RV8 nose was ditched by MGA…

…whose final design reverted to recessed lights

Early McGovern rendering sparked Rover changes

NEW MG

♦ Rovers, and becomes the first sports roadster in living memory with a gas-fluid suspension that connects the front and rear wheels.

Given the obvious care with which the MGF's shape and specification have been compiled, it's a surprise to learn that the MGF has only been on Rover's official model plan since 1992. For years, the idea of an all-new MG was sidelined by a management more interested in getting its mainstream models properly up to class standards.

Yet by 1992 engineers and designers had done a great deal of preparatory work. Several design consultants had finished full-size MGF concept models (see p18). Rover's styling staff had already gathered a number of traditional MG cars and reference material for what they called "The MG Event" to identify the qualities and hardware that they believed made a true MG. In the end, Rover Design decided to handle the project in-house, believing that the concepts lacked the elusive quality of 'Britishness'. They wanted, in particular, to give the body some of the 'tension' which distinguished the MG EX-E prototype shown at the Frankfurt motor show in 1985.

To strike the right balance of sporting character and practicality, MG engineers started by scanning dozens of magazine and newspaper road tests, believing that cars in the MGF class were widely bought as a result of what was written about them. They studied tests of the MR2 and MX-5, of course, but went back as far as the old MGB and even analysed comments about the most sporting of hot hatches, the Peugeot 205 GTi 1.9. Engine sound and response were highly valued, they found, along with crisp, faithful handling. Many tests emphasised a sports car's need to have a good touring range. And Rover's researchers were pleased to discover (in the interests of keeping insurance costs down) that top speeds and 0-60mph times were of relatively little importance.

But first they had to determine the MGF's true character, charting a way between the expectations of the flat-capped owners of old MGs and lovers of stripped-to-the-bone high-performance cars. The MGF, Rover says, is a sporting car in every way, but comfortable enough to cruise long distances (in a way an MX-5, for instance, is not) and be used as an owner's sole means of transport. It was also built to undercut both key Japanese rivals on service costs to 60,000 miles, through the use of components like long-life spark plugs and air filters.

Starting in 1989, three years before an MGF programme had been officially recognised, Rover's engineers built three prototypes — transverse mid-engined, front engine/front-wheel drive and front engine/rear drive. By 1992 they had already decided that the mid engine layout offered conclusive advantages in weight distribution and handling 'tuneability'. They built prototypes with all-steel suspensions, too, but concluded that a modified Metro

Engines

K-series too big for its boots? Try new boots

Rover used some ingenious engineering solutions to increase the size of its award-winning engine. Julian Rendell reports

The big-capacity K-series is an engine that should never have been built. Conceived as a family of small engines, the design squeezed so much into its compact dimensions that Rover persistently denied any chance of more capacity.

Yet six years after its first appearance, Rover has successfully stretched the all-alloy engine to 1.6 and even 1.8 litres. The MGF is the first to use the engine in both regular and variable valve timing forms.

"I'm delighted that what I said five years ago was wrong," says Sivert Hiljemark, engineering director of Rover Group Powertrain.

Driven by a need to replace its expensive, bought-in Honda 1.6, Rover spent "less than £200 million" and six years working on the big-block K, solving its problems with ingenious and innovative engineering.

The key features of the big block are new cylinder liners, called 'damp liners', which allow an increase in the cylinder size by squeezing bigger bores into the same overall block length. For both 1.6 and 1.8-litre engines that means 80mm bores (they're 75mm in the 1.4).

Rover's engineers have perfected the damp liners to such an extent that they rest on a lip only 1.4mm wide. The bore walls are only 3mm wide and the water cooling jacket around the liners is just 0.65mm. Despite the close tolerances, Rover is convinced that reliability, put to the test in more than two million miles and 50,000 test bed hours, is peerless.

The new bores are enough to boost capacity to 1.6 litres, but to get to 1.8 there's a new long-stroke crank with an 89.3mm throw.

Despite the internal alterations, the K's unique bolt-through sandwich construction, by which a single bolt holds cylinder head, block and sump, stays on unchanged.

Taking an opportunity to use the latest design techniques, the 1.8 crank has a larger diameter but shorter pins to which the big end caps/conrods are bolted, with 40 per cent more stiffness and 25 per cent more web stiffness. Courtesy of an alloy sump, block stiffness is increased threefold. Both engines have new, ultra-lightweight pistons that tip the scales at an extraordinary 194g each, nearly half the weight of the 1.4's pistons.

All this helps smoothness and refinement. "Normally bigger-capacity engines are less refined – typically second order forces go up five per cent – but we've managed to cut them by 12 per cent," says Derek Crabb, Rover's chief engineer for petrol engines.

Rover has pioneered 'damp liner' technology to squeeze extra capacity from K-series engine

Q&A
Alex Stephenson
Managing director,
Rover Group
Powertrain

Why go for variable valve timing?
We looked at everything, including turbos and superchargers, but VVC offered the best package of performance, fuel economy and exhaust emissions.

Why not use a turbocharger instead of increasing capacity?
We ditched the turbo because it would have given a peaky power delivery for such a tiny engine as a 1.4. Also, the turbo adds a lot of cost. On its own it's 25 per cent of the total engine cost, but with an intercooler it's one third. We also wanted to avoid any insurance and theft problems of a car plastered with turbo badges.

And a supercharger?
We watched sales of the Volkswagen G-Lader with great interest – and it never took off. It also adds a lot of expense.

What about reliability of the VVC?
The stresses involved are lower than those experienced every day in diesel engine parts like injector pumps, so it's not a problem.

Can you apply VVC to other engines – say, the forthcoming KV6?
That's in the 'yet to be dealt with' basket. One of the problems we must solve is setting up the drive on the vee engine. With three cylinders per bank, we need two hydraulic drive units and that's not cheap – or easy.

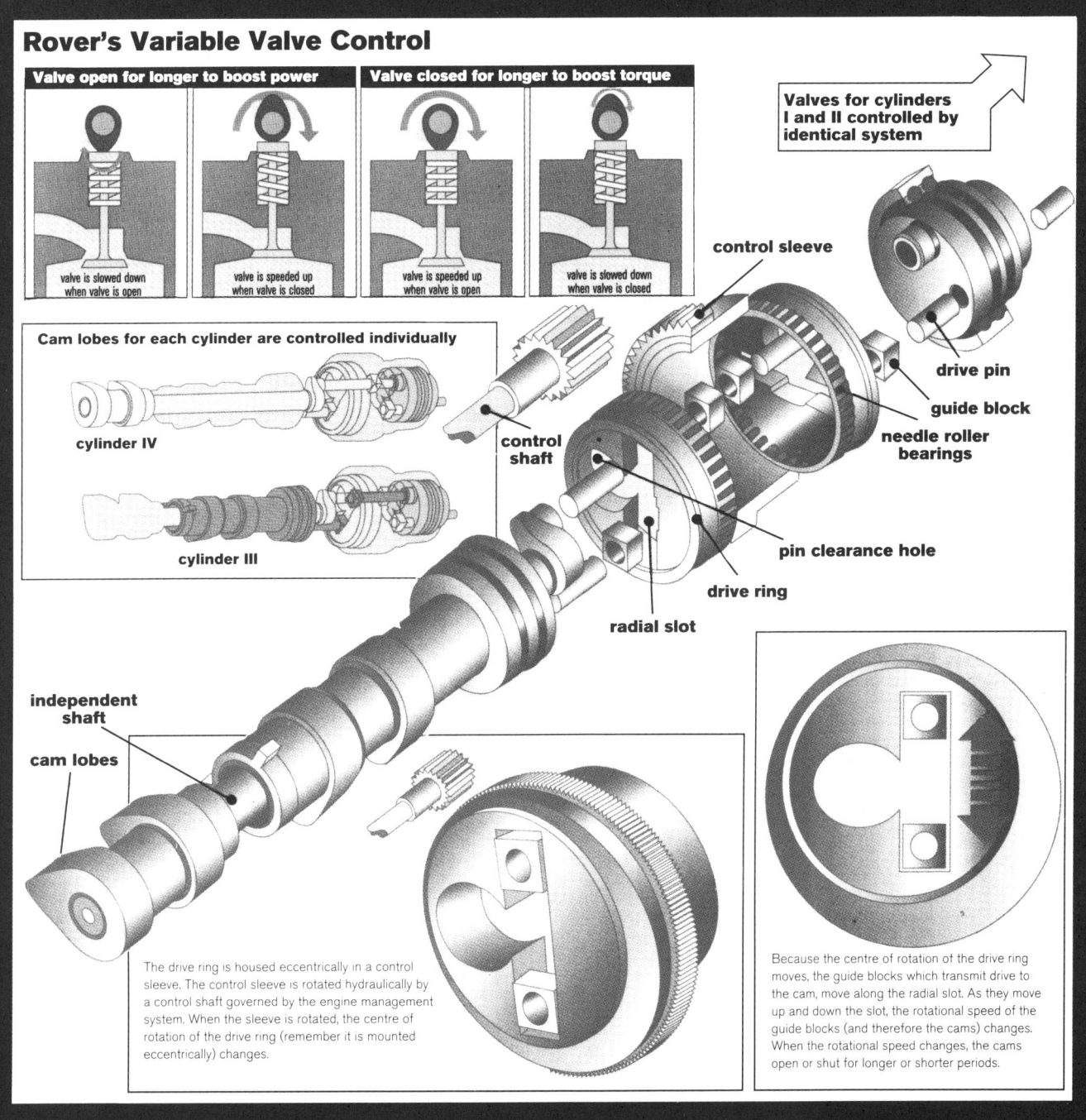

Rover's Variable Valve Control

Valve open for longer to boost power
- valve is slowed down when valve is open
- valve is speeded up when valve is closed

Valve closed for longer to boost torque
- valve is speeded up when valve is open
- valve is slowed down when valve is closed

Valves for cylinders I and II controlled by identical system

Cam lobes for each cylinder are controlled individually
- cylinder IV
- cylinder III

Labels: control sleeve, drive pin, guide block, needle roller bearings, control shaft, pin clearance hole, drive ring, radial slot, independent shaft, cam lobes

The drive ring is housed eccentrically in a control sleeve. The control sleeve is rotated hydraulically by a control shaft governed by the engine management system. When the sleeve is rotated, the centre of rotation of the drive ring (remember it is mounted eccentrically) changes.

Because the centre of rotation of the drive ring moves, the guide blocks which transmit drive to the cam, move along the radial slot. As they move up and down the slot, the rotational speed of the guide blocks (and therefore the cams) changes. When the rotational speed changes, the cams open or shut for longer or shorter periods.

K-series VVC links cams and drive mechanically, keeping inlet valves open for longer at engine speeds above 4000rpm. The result: 143bhp at 7000rpm

How variable valve timing works

Under the codename Hawk, Rover first researched variable valve control (VVC) for the K-series engine in 1989, but it wasn't until Easter 1993 that the first development engine ran. Tests quickly showed that Rover's design increased power and torque over a wide engine speed band, yet emissions and fuel economy were not adversely affected.

Given the go-ahead, Rover's engineers started to fine-tune their design, which was based on a lapsed patent held in the '70s by piston-maker AE. The result was a 1.8 engine with peak power of 143bhp at 7000rpm and maximum torque of 128lb ft at 4500rpm.

Described as 'third generation', Rover's VVC is technically ahead of BMW's cam phasing system and Honda's VTEC-E. Unlike cam phasing, which just alters valve overlap, and the Honda system, which switches between two different inlet cams, Rover's VVC can continuously vary the inlet cam period. It allows better combustion over a wider range of engine speeds, boosting power and torque.

The secret to VVC is the clever mechanical link between the inlet camshaft and its drive. An eccentric rotating disc, controlled by the engine management system, alters the relationship between the camshaft and the crankshaft. The effect is to keep the inlet valves open longer at more than 4000rpm, which, with bigger 9.5mm inlet valves, flows more air into the engine.

NEW MG

Low-profile pram-style hood, engineered by Pininfarina, folds away under tonneau. There's no power version, but there will be an optional hard-top

♦ Hydragas system suited a sporting car, perhaps even better than it did a cooking saloon.

Once the primary decisions had been made, the body structure was engineered by Rover and Mayflower's Coventry-based Motor Panels business. It has now begun making bodies to match the anticipated first-year production rate of 15,000 at Rover's Longbridge works in Birmingham, rising to 30,000 in subsequent years.

All along, the F-type was intended to make best possible use of existing Rover components, but engineers now insist that this was not a constraint in the key areas of suspension type and geometry, engine layout and cabin design. The F-type could have been front-engined and rear-wheel drive if that had been appropriate, they insist, even though Rover's saloons all have transverse powerplants. The car is built to last a long time in production, Rover insists, and all decisions on shape and specification have been taken in the light of that.

The MGF's dimensions, in every direction, are within a few of millimetres of those of its closest roadster rivals, the new Fiat Barchetta and five-year-old Mazda MX-5 — even though neither of those cars shares its transverse mid engine layout. And the MG weighs 1060kg at the kerb — almost exactly the same as the others.

The MG's body is a unitary, all-steel structure with a high degree of rigidity. Rover engineers estimate that its torsional strength equals that of a modern three-door hatchback, making it far more rigid than other convertibles and most specialist two-seaters on the market. The mid engine layout helps here, concentrating the mechanical mass in the strongest part of the car's structure.

Hydragas units are the springing medium for an all-independent suspension system which uses double wishbones at either end of the car. Unlike other Hydragas applications, the MGF system has both separate tubular dampers and an anti-roll bar, which Rover's handling experts claim provide the high degree of body control needed for a sports car.

It is the Hydragas system's interconnection, front to rear, that gives it conclusive advantages over rival set-ups, Rover says. The percentage difference in an MGF's weight, laden to unladen, is low compared with a small saloon's, and the system has allowed designers to make better use of the car's total wheel travel and improve ride.

There are two MGF models — a £16,000 car with a 118bhp version of the new 1.8-litre engine and a £18,500 model with variable valve timing and 143bhp (see p20). The torque outputs of the two engines are similar, giving rise to the thought that the lower-powered engine will suit drivers who depend most on an engine's mid-range power. The base 1.8 has 122lb ft at 3000rpm; the VVC unit produces only 6lb ft more at 4500rpm.

Both MGFs run on 6Jx15in alloy wheels with 185/55 VR15 tyres at the front and 205/50 VR15 at the rear. The VVC model has a new speed-sensitive power steering system which uses a micro-controlled electric motor to provide its servo effect (which reduces with speed) instead of the more complicated, more common Toyota MR2-style device of an electric motor powering a nose-mounted hydraulic system. The VVC also has anti-lock brakes as standard, which remain an option on the cheaper car.

Inside, designers have created a 'twin cockpit' theme, dividing the cabin

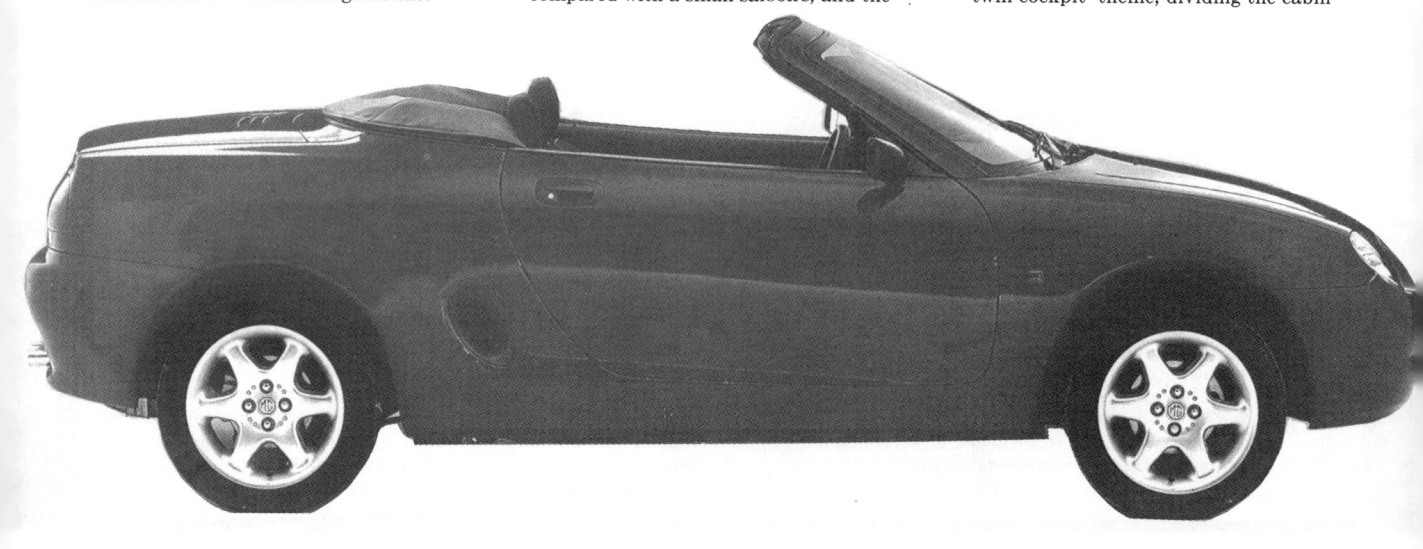

10

Suspension

Why the gas man is still laughing

The inventor of Hydragas suspension tells Steve Cropley why it's ideal for the MGF

Dr Alex Moulton says he's "rather pleased" that Rover has decided to use his Hydragas suspension system, now more than 30 years old, for its new MG sports car. But he isn't all that surprised.

Although critics of the system have forecast its demise for decades, Moulton believes it has conclusive advantages for a small roadster in compactness and low cost of manufacture. "The tooling was paid for many years ago," he says.

The best Hydragas property, Moulton says, is that it interconnects the car's front and rear suspension systems. "It de-fidgets the car," he says.

"In a little roadster there is a higher than usual potential for the car to become unsettled on rough surfaces, and the MG performs much better than the competition in that area, while continuing to handle and steer the way a sports car should."

Given that he has argued (successfully) in the past against separate dampers and anti-roll bars in Hydragas installations, Moulton is relaxed about the use of them this time. He still believes the system could be adapted so that it doesn't need them but admits that tuning the car's handling is easier if you have them.

He'll be driving his first final-spec MGF about the time you read this but has already heard from trusted sources within Rover that the car handles extremely well, while providing ride comfort beyond that of its competitors.

Moulton, an engineering consultant who was also responsible for a city bicycle which had its own rubber suspension, is bullish about the future. While some are already forecasting the MG application as the last for Hydragas, Moulton demurs. He believes his unique system has history on its side: nearly 10 million cars have been built with Hydragas suspension.

He has already managed to get BMW chairman Bernd Pischetsrieder to drive an "optimised" Hydrolastic Mini he built years ago, and stands ready to assist with the new Mini programme in any way he can.

"The future," he says, "is tantalising."

NEW MG

◀ with a prominent centre console and using substantial screen pillars and relatively high sides. They believe the move makes the car feel considerably more secure than rivals, while physically making the MGF safer. In fact, crashworthiness has been one of Rover's preoccupations; it claims the car exceeds all current and anticipated crash regulations easily, aided by the mid engine layout, which allows plenty of 'crush space' to cope with head-on impacts.

In common with the MGF's exterior, the cabin has few gimmicks, although the cream-faced instruments do hark back to the MGs of yore. But Rover has considerably fine-tuned the MG octagon badge, which appears on the car's nose and wheels, in a bid to give it modern impact while preserving a traditional look.

Both models have electric windows and what Rover describes as a "good quality" stereo. Two-level locking — incorporating a remote 'superlocking' system — goes into both models. They also feature a manually operated, pram-style soft-top, designed by Rover and engineered at Pininfarina. Although designers saw no need for an electric option, there is an optional hard-top.

Equipment unique to the VVC includes leather trim for the new MGF steering wheel, which incorporates a driver's airbag, and half-leather covering for the sports bucket seats.

The F-type will be launched in the summer — the first cars are expected to hit the road in June — but early buyers will only get the 118bhp 1.8i version. The higher output VVC model, whose engine is still being worked on, won't appear until later in the year, probably in October.

The MGF will be handled by 125 so-called 'specialist' dealers, who will be expected to take one car for stock and one demonstrator at launch. Rover is coy about its initial British volume but anticipates a lively demand for the car across Europe and in Japan as well as from its home market.

No details of a US debut have been announced, although it's worth bearing in mind that US demand made a success of all previous post-war MG models, and Rover could hardly ignore the country that has clamoured for a new MG roadster for so long.

All in all, it seems Britain has a new sporting flagship, and on first acquaintance, it seems to contain the right ingredients to meet the tough competition head-on and — more importantly — to please those legions of car lovers who have waited so long for a new MG. It's not quite so good as getting the Empire back, but nearly. ●

Sculpted, modern design maintains MG character

Boot bigger than you'd expect but soft bags only

12

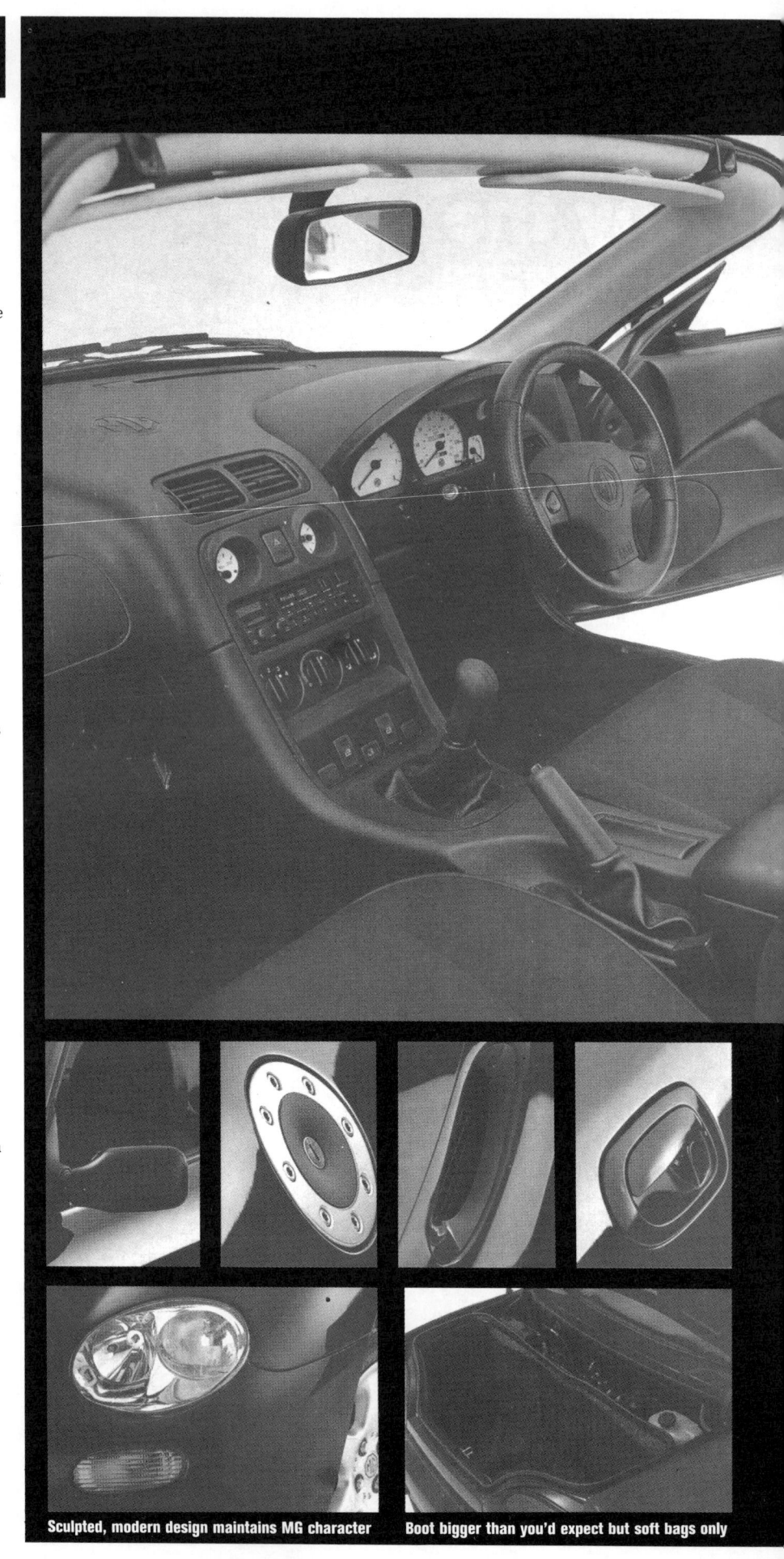

IN DETAIL

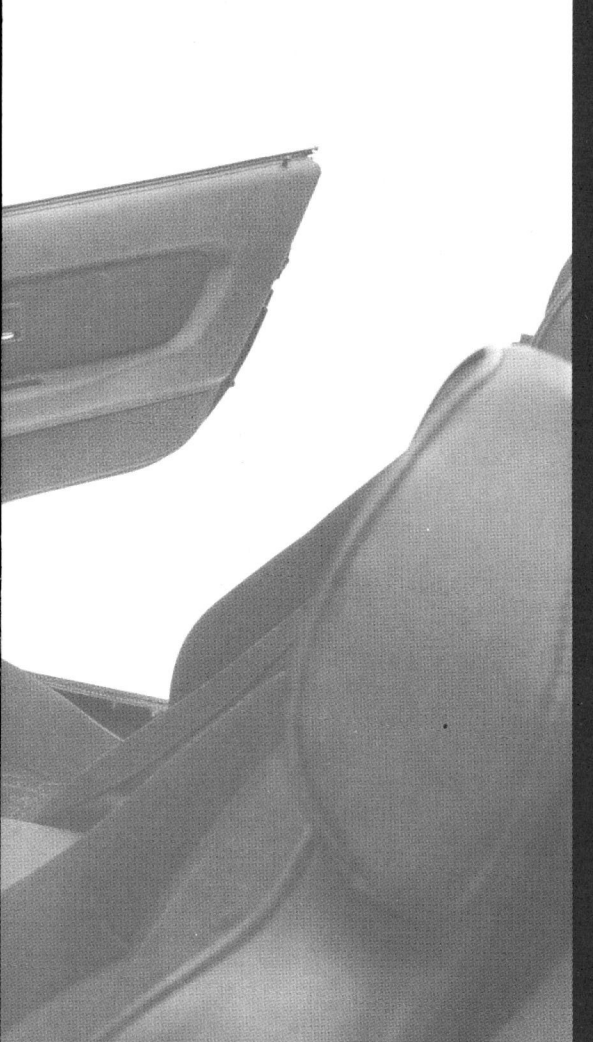

Cream dials echo old MGs, but retro look shunned

Smart new alloys have modified MG octagon

Clever packaging has enabled Rover to fit full-size spare. VVC-powered car has anti-lock brakes as standard

FACTFILE
**ROVER MGF
1.8i/1.8i VVC**

HOW FAST?
0-60mph	8.5/7.0sec
Top speed	120/130mph
MPG: urban	30.0/tbc
56mph	54.0/tbc
75mph	42.0/tbc

All manufacturer's claimed figures

HOW MUCH?
£16,000/£18,500
On sale in UK June/October

HOW BIG?
Length	3913mm (154.1in)
Width	1628mm (64.1in)
Height	1264mm (49.8in)
Wheelbase	2375mm (93.5in)
Weight (claimed)	1060/1070kg (2337/2359lb)

ENGINE
Max power
118bhp at 5500rpm/143bhp at 7000rpm
Max torque
122lb ft at 3000rpm/128lb ft at 4500rpm
Specific output 65bhp/79bhp per litre
Power to weight 111bhp/134bhp per tonne
Installation mid, transverse, rear-wheel drive
Capacity 1796cc, four cylinders in line
Made of alloy head and block
Bore/stroke 80/89mm
Compression ratio 10.5:1
Valves 4 per cylinder, dohc with hydraulic tappets
Ignition and fuel Rover's own Mems direct injection electronic ignition, multi-point sequential fuel injection, adaptive fuelling control
Gearbox type PG1-u, 5-speed manual

SUSPENSION
Front independent, double wishbones, Hydragas springs, anti-roll bar
Rear independent, double wishbones, Hydragas springs, anti-roll bar

STEERING
Type speed-sensitive electric power-assisted rack and pinion

BRAKES
Front discs **Rear** discs **Anti-lock** standard on VVC, optional on 1.8i

WHEELS AND TYRES
Size 6Jx15in **Made of** alloy
Tyres 185/55 VR15 (f), 205/50 VR15 (r)

SOLD BY
Rover Cars
PO Box 395
Longbridge
Birmingham B31 2TB
Tel: 0800 620820

road test

ROVER MGF 1.8

MaGic!

road test

Is the MGF Britain's best new sports car?

WORDS BY RICHARD MEADEN   PHOTOGRAPHY BY DOMINIC FRASER

Cut me an extra large slice of humble pie, Mrs Miggins – the MGF is absolutely fantastic. Thing is, I didn't think it would be. Disappointed by uninspiring early photos of it and considering Rover's track record for building dynamically exciting cars, the outlook for the first proper MG in almost 20 years didn't seem too bright.

A mid-mounted 16v engine, rear-wheel drive, pert soft-top styling and an evocative badge are juicy ingredients alright, but we're talking Rover, remember. 'It won't drive as well as it looks', I thought, and as for trumping the MX-5 and Fiat Barchetta, *pah*, no chance mate!

Well my scepticism was unfounded, ill-judged and just plain bloody wrong. Yep, the MGF is more gorgeous than a very gorgeous thing indeed. For a start, it's *soooo* small. Cutely compact, well proportioned and just so right that you can't help but *oooh* and *aahh* at its cherubic curves.

road test

Shying away from the blatant 'Hello boys' buxomness of Fiat's Barchetta, the F is more a collection of subtle curves, classy swoops and neat touches that combine to form a smooth, wedgy lozenge on wheels. It's modern, but retains enough retro charm to tempt born again boy racers. In fact, whether you're wearing a baseball hat or a flat tweed cap, you'll look at home in the MGF, and that's no mean feat.

Good news so far, then, but what about price? At first glance £15,995 for the MGF 1.8i (tested here) and £17,995 for the more potent 1.8VVC seem a bit steep, but look closely at what you get for your dosh and the balance tips firmly in the F's favour. A Fiat Barchetta is around £14,000 but its steering wheel's on the wrong side, it has a rubber mat on the floor instead of carpet, and it does without central locking. Likewise the basic Mazda MX-5 1.8i (£14,495) – cheaper than the MG it may be, but its trim level is barely 'Popular Plus', lacking even power steering.

OK, so I've eaten a generous wedge of humble pie, my hat and one leg of my boxer shorts already, and I haven't even driven the MGF yet. Better have a look inside before I clear the plates away and wash up.

AT THE WHEEL

Damn! Looks like I'll be having seconds. Swing open the big driver's door, slip behind the small, stylish but sadly plasticky steering wheel (why no leather rim?), and already I know I'm going to like this car. It feels tight, together, well-sorted, focused. It's comfortable and snug, like a made-to-measure suit. The driving position's less sporty than expected, higher and slightly upright, while the seat and steering wheel are smaller than normal, so you feel one size too big for the car. But this 'honey, I shrunk the MG' sensation soon wears off, and after 10 minutes it fits like your favourite jeans.

Neat touches abound. The restful cream-faced dials with burgundy octagons look good and disguise their Rover origins well, while the embossed MG badge on the dashboard breaks up what would otherwise be a bland expanse of plastic. The ergonomics are excellent too, with thoughtfully placed switches and clear instruments. Everything has a quality feel. The pedals are well-positioned for heel and toeing, while their weight and consistent action makes for smooth, effortless progress. The gearshift is similarly clean and slick, if a touch vague, but this is only a very minor gripe.

The overriding impression is that the MGF is a car built to last. Nothing feels weak or fragile, and one senses that Rover has done millions of development miles, so well-sorted is it.

It's always tempting in a soft-top car to whip the roof down at every opportunity, when a much sterner test is to drive it with the roof erected. That way the *al fresco* sounds of summer and the bluster of a warm breeze can't hide chattering, rattling or squeaking trim. And the MGF is miraculously solid roof-up, unlike, say, the Barchetta, which has more rattles and squeaks than a branch of Mothercare. Unclip the two header rail fasteners, flip the Pininfarina-engineered mohair hood down, and the next surprise is the lack of buffeting in the cockpit. Windows up or down, the slipstream leaves you unflustered, even on the motorway. All you get is a slight ruffling around the back of your neck. Clearly there are some very clever (or very flukey) aerodynamics involved here. Whatever, it makes the MGF one of the most comfortable top-down cars you can drive.

PERFORMANCE

Without the all-singing and dancing valvegear of the sportier 143bhp VVC model, the cooking 1.8-litre MG was never going to set tarmac alight. Fortunately, it still makes excellent use of its 118bhp.

The enlarged K-series engine is a sweet motor, free-revving, smooth, refined and quiet. It's also gutsy from low revs and although it never really seems to hit a power peak, the evenly spread shove is effective enough.

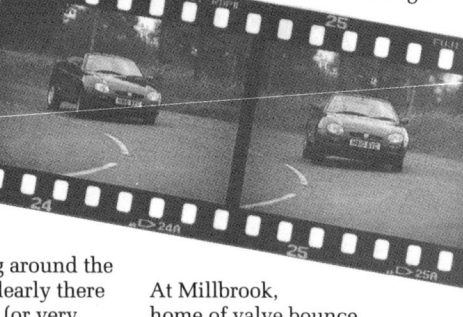

At Millbrook, home of valve bounce, wheelspin and tyre smoke, it takes a temporary suspension of mechanical sympathy and a few *bababababa* rev limiter standing starts to hit 60mph in 8.6secs, a

'F' represents a new dynamic high for both MG and the Rover group. Finest handling MG ever

'The MGF looks modern, but has enough retro charm to tempt born again boy racers'

tenth down on Rover's claimed time. In-gear times illustrate the healthy mid-range urge and flexibility of the engine, although acceleration tails off markedly above 100mph. On the high speed bowl in fifth, the engine slowly manages to haul itself up to peak power revs, building slowly but surely to an indicated 130mph, (a true speed of 121mph).

The K-series has always been a modest drinker and although the engine has grown, its

Rear-wheel drive, mid-engined – clearly, the MGF offers scope for power oversteer. And it's fab

BUT IT'S A HAIRDRESSER'S CAR...

It's one of our most treasured traditions – when you see a sportscar you know you can't afford, being driven by a beautiful person who hasn't a care in the world, you must mutter under your breath: 'hairdresser's car'.

The MGF is a prime target – in the months to come, as it's driven casually through the streets and parked smoothly outside trendy coffee bars, it'll be dismissed as a car for people with combs in their breast pockets. So we thought we'd conduct a scientific test, and see what the hairdressers themselves think.

Toni & Guy's is a trendy 'salon' in Cambridge. We parked the MG right outside and asked the 'stylists' to come out into the sunshine to give us an opinion.

Stuart, the manager, drives a Saab, though he's thinking of buying a 1974 Alfa Romeo Spider. 'I love it' he says, looking along the F's flanks, nodding slowly. '£16,000 huh? Yeah, put me down for one. That's a lot of money, but I'd buy it because it's gonna be a cult car'.

Tara and Natalie come out next, to sit in the car, soak up the admiring glances and Japanese flashbulbs. Tara drives an XR2i: 'I like this a lot,' she says, fondling the fitments, 'if I had 16 grand, I'd buy one'. Natalie doesn't own a car, but she's got an eye for details: 'It's a nice shape, and I like that *(pointing to the bare metal fuel filler with its counter-sunk socket cap screws)* – is that where you put the petrol in?' Yup. I ask her if it would bother her, having a convertible. Wouldn't you be worried about your hairstyle, Natalie? 'Not with my fuzzy mop,' she replies with a laugh. Earthy, unpretentious. I like it.

But it's Myeesha who offers the killer blow (dry). 'Men drive Saabs and BMWs. This would make a great lady's car,' she says. Eeekk!

Mark Walton

Stuart – 'It's gonna be a cult car'

Tara & Natalie – happy bunnies

And then we asked a punter...

'Bugger off, I'm busy'

road test

'The MGF is so friendly and exploitable – it never feels snappy or sharp'

road test

MGF has all the character of its forebears (like the MGA pictured here), but a much wider handling envelope. It's a proper sports car

F looks smaller in the flesh; has cute curves

thirst hasn't. Rover claims the cooking 1.8-litre will take you 33.8 urban miles on a gallon, or 39.5mpg with mixed use. Clearly they didn't use the *Performance Car* team to do the driving, as we averaged closer to 30mpg during our loan period. Still, with the combined weight of our right feet born in mind, this bodes well for 'real' MG owners.

On the road, the MGF makes a willing and able B-road companion. Working through the well-judged ratios, spirited progress requires seemingly little effort. The engine sings enthusiastically behind your shoulders and the chassis encourages you to push through twists and turns, preserving momentum. Second gear is seldom required unless the corner is particularly tight, or you fancy a bit of tail-out fun.

Come up against a car travelling at 50mph on a typical duck-out-and-have-a-look B-road, though, and the engine's lack of genuine zest becomes apparent. It makes overtaking a planned manoeuvre rather than a spontaneous act.

What you lose in outright, high rev response, you gain in mature, adult performance. You dictate the pace and tempo of the journey, rather than suffering the sub-5000rpm slumber of a more frantic 16v engine. You don't have to rev the rods off the 1.8 K-series engine but should you feel the need for large revs, it ups the ante nicely while remaining as refined and polished as ever. Only engine character is in short supply, with little in the way of rorty, snorty noises.

The great thing about this MG is that it's a complete package with no weak spots. On the road and at Millbrook the anchors feel strong, doggedly wiping off speed without drama. Balance is what the mid-engined F is all about, and it wrings every last ounce of stopping power from the vented front and solid rear discs. After 10 pad-pounding, disc-warping fade tests from 83mph, smoke is whisping from all four wheels, but the stoppers have barely wilted. Barker, a serial brake killer, has witnessed nothing like it before and is deeply impressed.

HANDLING

If you've driven a Toyota MR2, you'll know that a mid-engine/rear-drive layout doesn't necessarily mean driving thrills by the bucket-load. Barker voiced all our concerns for the MG's chassis when he said, 'please let the dynamics be good, not just safe enough so you can recommend it to posey women, but sporty, blokey good – responsive, adjustable, communicative fun'.

Well praise be to Rover, because the MGF is all this and more. Driving at a normal pace initially, it's hard to say whether it's front or rear-drive, or indeed front, mid or rear-engined. There's no scrabbling from the front end, or nervous twitchiness from the rear, just solid, consistent and weighty feel from the optional electric power-assisted steering.

Once acclimatised to the controls, upping the pace reveals the unusual depth of

SECOND OPINIONS

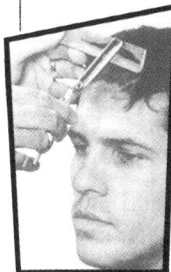

Mark's fringe was cut using thinning scissors to add texture

I didn't think I'd like the MGF. Sorry Rover, it's nothing personal, but when the first wave of publicity washed over the nation earlier this year there was something about it that made me feel it was going to be a cop-out, not a real sportscar. A girly, white-Escort-XR3i-Cabriolet kind of thing. And I was wrong. The MGF's internal dimensions and driving position are more like a hot hatch than an MX-5, and the steering lacks some sharpness, but the package as a whole is well-sorted and serious fun. Its mid engine isn't just a gimmick – through roundabouts and long curves the F has a natural balance, tempered by a progression and predictability that suggests hours and hours of patient development work. The MGF inspires confidence, reacts well to your inputs and comes alive when you drive it harder. I was wrong; I like it mucho.

MARK WALTON

John opted for the fringed feather-cut, with mousse for extra body

The camera does lie. I was expecting a bigger car but in the flesh the MGF is a compact, pert design. After strolling around it twice, I just wanted to get in and drive it away. It was raining, but I enjoyed myself so much at the first roundabout, I did two laps. You need to know what you're doing, though. In the dry the MG handles like a front-driver but in the wet, you certainly know it's the rear wheels that are driven. It's still one of the most controllable mid-engined cars I've played with and easily the best Rover I've ever driven.

JOHN BARKER

road test

the F's talents. It really does have terrific balance and poise. Attack a corner in conventional front-drive manner, running deep into it, testing the front end grip, and the F bites on turn-in like a good hot hatch and hacks on through neatly. 'But it's rear-wheel drive!' I hear you cry, 'where's the oversteer?' Fear not, for the MGF offers understeer, neutral balance, lift-off oversteer and power oversteer on demand.

Change your cornering technique and the F adapts with you. Lift off sharply mid-corner, and the tail gently slips round, cancelling any understeer, tightening your line and pointing you straight for the next section of road. Re-apply the power as the tail begins to slide, and you can hold the F in a delicious power-slide, seemingly for as long as you like. It's so friendly and exploitable, and never feels snappy or sharp.

Sprinkle some rain on the tarmac and the MG resists morphing into a whirling dervish, but you do have to be quicker on the draw when it comes to corrective steering inputs. In such conditions, the slowish steering becomes more of a problem, although its consistency and feedback will still see you through. Our only concern is that those who don't push hard in the dry will expect the MG to retain its surefooted front-drive-*esque* handling balance in the wet, only to discover that lifting off mid-bend or hoofing it at a roundabout sets the tail swinging. It's far from treacherous, but you do need to be on the ball.

The really astounding thing about the F is it's unshakeable composure, whether you're working the chassis really hard through a corner, or pounding along badly pot-holed roads. The Hydragas springs (interconnected front to rear) and double wishbone suspension copes with badly mettled tarmac and fast undulations with equal aplomb, controlling body movement with an iron fist, but riding the rough stuff with the smooth caress of a velvet glove. A Mazda MX-5 feels like an ox-cart in comparison.

VERDICT

No point beating about the bush. The MGF is a winner. With its wonderful styling, entertaining chassis, exceptional ride and rock solid build quality, the F mercilessly exposes the shortcomings of the ageing MX-5 and the toy-like build quality of the Barchetta. The only drawback is face-ache from smiling so much, and you can't ask for much more than that from a 16K sports car. It's a great car and one Rover can truly be proud of because it's their own, not a re-grilled Honda or a BMW in disguise. Better still, it's a proper MG, not a tin-box with octogans. Our only worry is there won't be enough to go around.

SPECIFICATION MG

Engine	Four-cylinder, in-line
Location	Mid, transverse
Displacement	1796cc
Bore x stroke	80.0mm x 89.3mm
Compression ratio	10.5 to 1
Cylinder block	Aluminium alloy
Cylinder head	Aluminium alloy, dohc, four valves per cylinder
Fuel and ignition	Electronic multi-point fuel injection and ignition
Max power	118bhp @ 5500rpm
Max torque	122lb ft @ 3000rpm
Transmission	Five-speed manual, rear-drive
Front suspension	MacPherson struts, lower wishbones and anti-roll bar
Rear suspension	Independent by multilink, coil springs and anti-roll bar
Steering	Rack and pinion, electric power assistance (optional)
Brakes	Front vented discs, rear discs. Anti-lock (optional)
Wheels	6.0J x 15in alloy
Tyres	185/55 VR15 (front) 205/50 VR15 (rear) Goodyear Eagle Touring
Wheelbase	7ft 9in
Fuel tank capacity	11 gal / 50 litres
Weight (kerb/test)	2337lb / 2667lb
Power-to-weight	99bhp per ton
Basic price*	£15,995
Alloy wheels	Standard
Traction control	n/a
Auto gearbox	n/a
Airbag, driver/pass	Standard/£345
Power steering (elec)	£550
Alarm/immobiliser	Standard
Anti-lock brakes	£650
Adj. steering col	n/a
Hardtop	£995
Leather trim	n/a
Metallic paint	£230
Price as tested	£17,770 Metallic paint £230, passenger airbag £345, power steering £550, anti-lock brakes £650
Insurance group	12 *(blimey!)*

PERFORMANCE

ACCELERATION THROUGH THE GEARS (SECS)

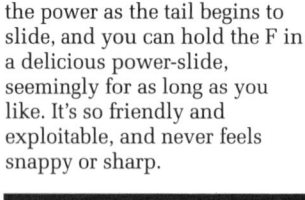

0-30mph	2.8
0-40mph	4.3
0-50mph	6.2
0-60mph	**8.6**
0-70mph	11.7
0-80mph	15.2
0-90mph	20.5
0-100mph	27.1

Standing 1/4 mile (secs/mph) 16.6/83

TED (secs/ft) 6.5/560
(Time Exposed to Danger) Time and distance required to overtake an articulated lorry travelling at a constant 45mph

Top speed 121mph

ACCELERATION (SECS) 3rd 4th 5th

20-40mph	5.4/7.4/11.4
30-50mph	5.3/7.0/11.3
40-60mph	5.3/7.2/11.3
50-70mph	5.8/7.7/11.5
60-80mph	7.0/8.1/13.2
70-90mph	12.4/9.4/15.8
80-100mph	- /12.6/18.1
Speed per 1000rpm in top	22.1mph

BRAKE TEST: Distance to stop from 30mph, 50mph and 70mph

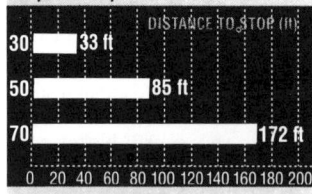

- 30: 33 ft
- 50: 85 ft
- 70: 172 ft

FADE TEST: Pedal pressure needed to stop at 0.5mg retardation from the quarter mile speed (TEN STOPS FROM 81 MPH)

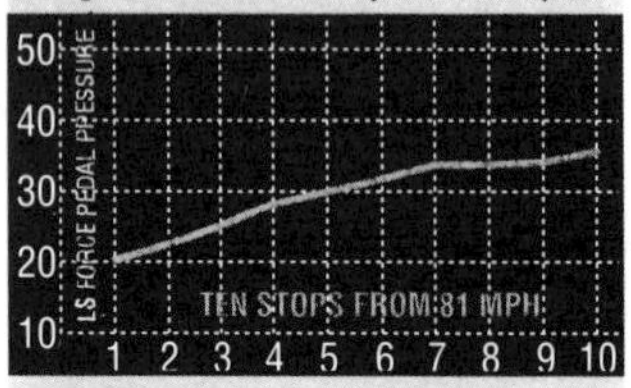

Test mpg	28.6mpg
Touring mpg (from Govt figs)	39.5mpg
Track conditions	Dry
Temperature (C)	14
Wind speed (mph)	12
Pressure (mbar)	995

Cabin is snug, well-screwed together. Nasty wheel, though

F IS FOR FABULOUS

*The **MG** nameplate has **returned from the grave;** Ian Adcock reports on the **MGF, the first** all-new car to wear the Red Octagon **since 1962.***

It's been a long wait—14 years and five months, to be precise. Particularly since Rover teased us with the MGRV8 in 1992, however, MG enthusiasts the world over have been counting down the months, weeks, days, hours, minutes and seconds until their long-gone marque would make its grand re-emergence as a genuine competitor in the sports-car market. The wait, at last, is over.

Here it is: The all-new MGF is a taut, muscular, mid-engine 2-seater that blends scads of MG tradition with wholly up-to-date technology and a goodly dose of value. It's a car that makes the Miata MX5 look passé in comparison and throws the styling gauntlet down to Fiat's cute new Barchetta. No doubt it must also be sending shivers down the spines of Rover's American-based BMW cousins, whose own upcoming Z3 Roadster may someday find its sales prospects being poached by cars wearing the traditional Red Octagon.

As with the BMW Z3 and the high-market Porsche Boxster, only the mechanical details of the MGF are available for now. Long-lead road testing had yet to take place

and nostalgically ape bygone MGs, but happily they've resisted taking the pure-nostalgia course. Instead, what you see here is a car that's as fresh and exciting as any of its predecessors were in their day, yet immediately recognizable as a member of the MG family. Certainly there's some homage paid to the MGB around the grille, right down to the criss-cross grating and the double creases above the Octagon, but that's about it apart from the car's discreet and tasteful wheel and trunk badges. There are some very nice touches, too, like the exposed filler cap, the 1960-style headlight covers and the fully functional side vents.

Inside the cockpit the story is much the same—the MGF remains an amalgam of the best of old and new. The twin-cockpit layout with its high-waisted doors looks instantly comfortable and buffet-free, while the driver has a neatly arranged console directly ahead featuring cream-colored instrumentation and a traditional 3-spoke, airbag-equipped steering wheel. MG aficionados will note with pleasure the red piping and fluting on the seats, both traditional MG styling cues.

Unlike Fiat's Barchetta, the one-piece Pininfarina-designed top doesn't disappear totally beneath the rear deck, standing

as SCI was putting this issue to bed, but the final shape and specification are now known at last.

Remarkably, the low-dollar MGF promises to go as well as it looks. Two new 1.8-liter fuel-injected versions of the Rover K-series Four will debut in the MG: The standard 1.8i will have 118 bhp @ 5500 rpm and 122 lbs.-ft. of torque at a low 3000 rpm. This will give a top speed of about 120 mph and 0-60 blasts in the mid-8s. Above that model is a Variable Valve Control (VVC) version boasting 143 bhp @ 7000rpm and 128 lbs.-ft. @ 4500rpm—enough power, Rover says, for 130 mph and a 0-60 run in seven seconds flat. Neither set of figures sounds at all optimistic for a car weighing in at around 2350 pounds, so we can reasonably expect production versions to better those numbers. The benchmark Miata does 0-60 in 8.9 seconds.

Both versions drive their rear wheels through a 5-speed transaxle, and 4-wheel disc brakes are fit all around, with ABS coming standard on the 1.8i VVC model.

It would have been simple for Gordon Sked's styling team to take the easy route

Busting Skinners Union: *The F-type's Acura-esque taillights hint at the strong relationship Honda shared with Rover before BMW bought its way in. Speed-variable valve actuation, double wishbones and fuel injection replace the mechanical tappets, live rear axle and twin SUs of MG's <u>last</u> new sports car.*

instead a little bit proud just as the tops did on the MGA and MGB. There's also a 1-piece fiberglass hardtop option complete with heated rear window. Similar to the newest entries from Fiat, Alfa, BMW and Porsche, MG has opted for a stout rollbar built into the windshield surround rather than a separate Targa-style roll hoop.

It would also have been easy for Rover to dive into the parts bins and base this new MG on a front-wheel-drive Rover 200 or 400 floorpan, but their admirable decision to opt for a mid-engine layout and double wishbone suspension all around prevented that.

Going to a mid-engine design brings a host of performance and usability tradeoffs, and every powertrain layout has its fans and foes. The mid-engine scheme also, however, makes using an existing front-engine/front-wheel-drive powertrain in this new location relatively straightforward; one simply moves the whole lot four or five feet back and adjusts the suspension mounts accordingly, resulting in relatively lower development and per-unit costs than a front-engine, rear-drive configuration such as Mazda selected for the Miata.

The MG's rather more advanced suspension medium promises much, however: double wishbones and antiroll bars front and rear should help to make its handling best-in-class, while a refined version of the Rover's well-proven Hydragas damping system should promote an excellent ride.

Equally novel in MG sports-car use is a new electric speed- and load-sensitive power assisted steering system. (This comes standard on the top-range 1.8i VVC and is optional on the 1.8i.) The steering effort required for any maneuver is intuited by a torque sensor in the steering column, which transmits this information to the ECU; that unit then compares the vehicle and engine speeds and calculates an ideal level of power assistance, which is finally made available back at the steering box.

While Rover has been generous in developing new technology for this car, it has sensibly saved costs where it can, especially in body assembly. The firm generated the body's actual design along with its crashworthiness figures, and also developed the body tools for its construction in-house. The actual task of turning the body from an engineering concept into a usable platform, however, was undertaken by subcontractor Motor Panels, which will also fabricate the finished shells. Painting, trimming and final assembly will be carried out at Rover's Longbridge, Birmingham, factory.

Sun, Surf and England: *Amazingly, the denizens of Europe's drizzliest island never quite did figure out how to make a leakproof convertible top in their heyday. This time around they've turned to the experts at Pininfarina, whose earlier Alfa Spider designs used to be the only cozy ragtops on the Continent.*

Rover's management fully appreciates they have something precious in the MG badge, and that "feel-good factor" will initially be preserved by sending the MGF out only to a highly specialized dealer network. Only 125 of the UK's 700-odd Rover dealers, for example, will be offered a shot at the new MG franchise.

In addition to its heritage of performance, however, one of MG's most significant attributes has also historically been its affordable pricing. While the large number of outsourced components and relatively low production numbers of the MGF will prevent it from competing head-to-head with cars like the Fiat Barchetta and Mazda Miata, its $24,250+ base price can hardly be called extravagant, either. Even fully loaded with VVC, ABS and an optional hardtop, the MGF buyer should be hard-pressed to spend $30,000 for this latest English offering.

Start saving and join the queue; almost certainly, the F is only the first example of the all-new MGs in our future. ●

THE AUTOCAR ROAD TEST NUMBER 4152

MGF 1.8i

This is the first proper MG for a generation with both the specification and looks to take Britain's most popular sportscar marque back to its glory days

If it is surprising that we have waited over thirty years for this, the next new MG, it is altogether more amazing that it is here at all. Since the MGB appeared in 1962, the marque has had to survive on a diet of increasingly old and unappetising product, suffer the indignity of having its name poached and plastered across a range of largely undeserving cars and, in between, long periods on the shelf. Most marques would have given up the struggle.

Yet still the name retains its magic. Were this not the case, the MGB would never have been disturbed let alone exhumed, reworked and sold as recently as last year as the dynamically challenged RV8.

The thinking is simple. If the RV8, obsolete twenty years before it was built, can sell then a proper, state of the art MG sportster should have them queuing around the corner.

But the MGF is more than that. It is also the first all-Rover car since the launch of the Montego in 1984. As such it stands or falls on its own merits.

At first glance, those merits seem compelling. Beautiful in photographs, it's close to stunning in the flesh, making the Fiat Barchetta look contrived and the Mazda MX-5 seem old. And unlike these principle rivals, its design is genuinely innovative. Where the Fiat uses a hatchback-derived front drive layout and the Mazda a traditional front engine, rear drive arrangement, the MG places its 1.8-litre twin cam 16-valve engine directly behind its driver making it the world's first affordable mid-engined convertible.

With that name, those looks and a basic list price of £15,995, its potential is clear. Question is, does it live up to it?

Performance
★★★
★★

There are two versions of Rover's 1796cc K-series engine for the MGF. The standard version is seen

LIST PRICE £15,995 **TOP SPEED** 123mph
0-60MPH 8.7sec **30-70MPH** 8.7sec **MPG** 26.4

FOR Beautiful and clever styling, engineering integrity, huge grip, superb brakes, exceptional build quality
AGAINST Considerably less fun to drive than an MX-5, dowdy looking cabin, compromised driving position

THE AUTOCAR ROAD TEST

here with 118bhp at 5500rpm and 122lb ft of torque at 3000rpm. An extra £2000 buys the VVC version (variable valve control) pushing the outputs up to 143bhp at 7000rpm and 128lb ft of torque at 4500rpm.

In standard tune, the MGF feels reasonably swift. The engine is as smooth as we now expect from the K-series but, thanks to its longer stroke, it has an even spread of torque meaning easier progress and fewer gearchanges. There is, if anything, not quite enough drama in its progress meaning the MG feels slower than it is. An MX-5, slower on paper, feels faster on the road.

Even so, few will find fault with the MG's 0-60mph time of 8.7sec even if that figure is aided somewhat by the natural traction advantage of having the engine's weight on the rear wheels. All out it tops the 120mph claimed by Rover by 3mph, auguring well for the 130mph claimed for the swifter VVC variant.

In-gear performance is solid but hardly exhilarating. It requires 12.1sec to cover the 50-70mph increment in top, enough to earn your attention but still hardly spellbinding stuff by the standards of many cheaper hot hatches.

Here lies the first clue to its overall purpose. It is an overwhelmingly calm performer, feeling no need to dress its acceleration in gaudy engine noise, or hurl its driver through gear ratios stacked like sardines in its five-speed box. Though the change quality is quite exceptional by all standards bar those of the MX-5, its ratios are widely spaced and relaxed, promoting refinement and economy but reducing in-car entertainment.

Economy
★★★ It should be no surprise that with its light-weight, aerodynamic shape, long gears and efficient engine, that the MGF is a reasonably frugal sports car.

During testing it returned 26.4mpg, rising to 33.8mpg on a leisurely motorway run. Most disappointing is the fuel tank which, at a miserly 50 litres, means each tankful rarely lasts more than 280miles.

Handling
★★★★ A mid-mounted engine with double wishbones at each corner is a layout more usually associated with Ferraris than affordably cute convertibles. Add in tyre sizes that vary from 205/50 VR15 at the back to 185/55 VR15 up front and Alex Moulton's brilliant Hydragas springing, and the potential of the MGF's chassis is obvious. But by no means will everyone feel that it has been realised to the full.

Rover has clearly decided to dedicate the MGF's chassis talent to covering the ground as swiftly and with as little fuss as possible. That this means true driver involvement has been seriously compromised will be seen as a crying shame for those hoping for a car with the same sense of fun as the MX-5.

The MGF is nothing like as amusing to drive as the MX-5. But there's no denying the MG, in its chosen role, is supremely effective. It will corner easily at speeds which would have the Mazda skittering nervously. Its body control is top drawer and even at its limit it remains absolutely faithful in its responses. No other car with

MARK BLUNDELL ON THE NEW MGF

'An approach was made because Rover wanted someone in the car to give feedback from a different angle. I haven't done any road car work before, my experience has been in the quickest things there are.

'I did one day's driving at the Gaydon test circuit, trying different configurations.

'I came back and spoke to the engineering staff. We had a small meeting and I also spoke to a couple of other guys who had been testing the vehicle.

'I came across as a race driver, but also took the view that the car is not just for the guy who thinks he can race the pants off somebody. It is also for the driver who needs to think about parking and shopping.

'And I said "It is a bit too much like this, it is a bit too like that…". In some areas, I was already on the right track as the Rover guys were thinking on these lines. I think I just put a bit of top spin on those thoughts…

'I gave them some pointers to which they were very receptive. It gave them something to think about and I think it allowed them to make a step forward with the car.

'For me, the MGF is a good, solid, sportscar. We did comparison tests with other manufacturer's vehicles – they were not in the same league.

'It can be pushed to the limit and not give the driver a fright. Predominantly, you have under rather than oversteer.

'I feel, for the average driver, that this is much safer because you can back off the throttle and bring in the front end, as opposed to trying to use a lot of ability to catch it.

'The variable valve timing engine was the hot ticket. The standard engine is very good, nippy and responsive, but for anybody who is looking for outright performance, the VVC engine is best. That was the one that gave me the biggest buzz.'

Hood does little to spoil smart lines. Optional hard top costs a cool £995

Badge is back where it belongs

Remote central locking is standard

Racy looking filler is a nice touch

Hood mechanism is a doddle to use

26

IN DETAIL

Cabin is spacious but disappointingly drab. Rover missed an opportunity to carry the style of its exterior into the cockpit. Instrument dials help, though

Boot space excellent, engine access poor. Very little extra room in nose

Gorgeous alloys are standard and well protected with lockable nuts

THE AUTOCAR ROAD TEST — MGF 1.8i

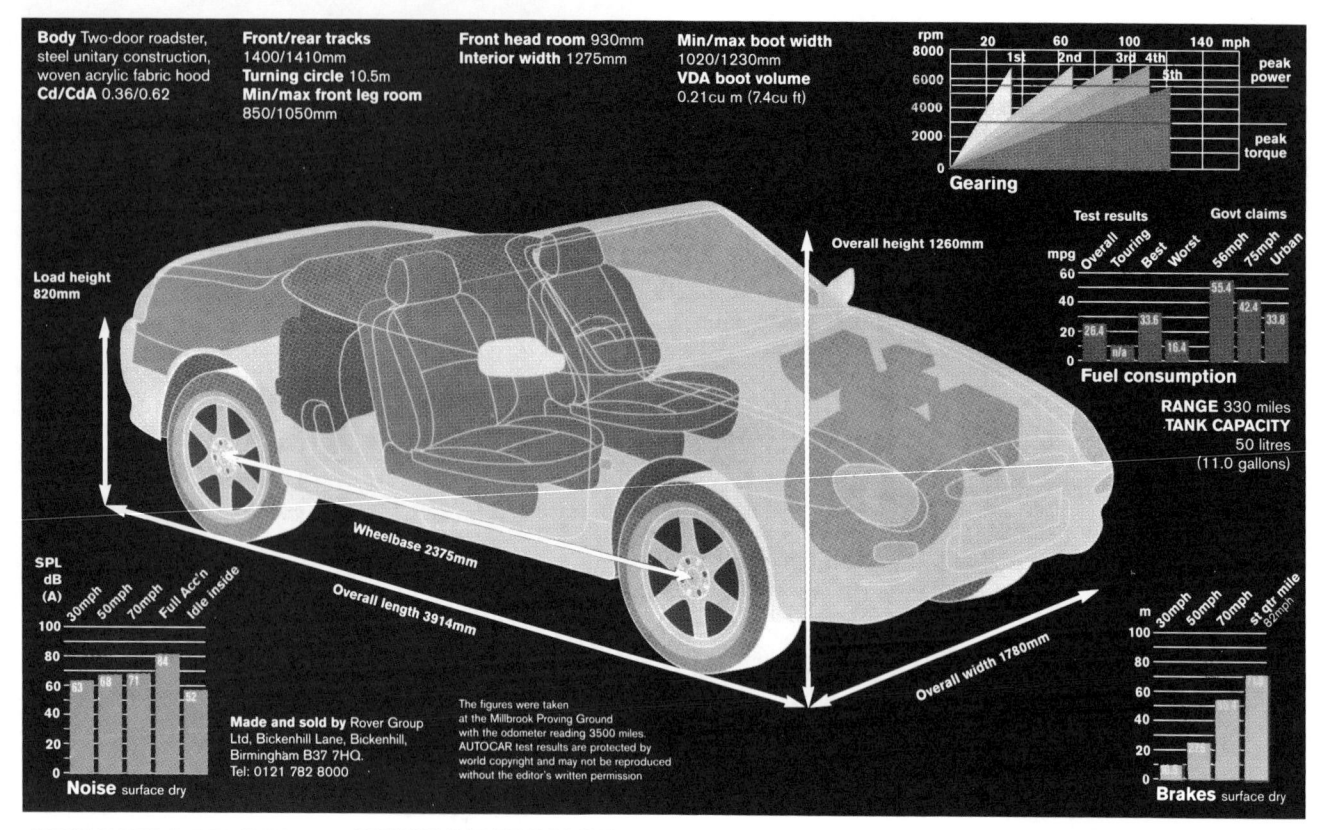

Body Two-door roadster, steel unitary construction, woven acrylic fabric hood
Cd/CdA 0.36/0.62
Front/rear tracks 1400/1410mm
Turning circle 10.5m
Min/max front leg room 850/1050mm
Front head room 930mm
Interior width 1275mm
Min/max boot width 1020/1230mm
VDA boot volume 0.21cu m (7.4cu ft)
Overall height 1260mm
Load height 820mm
Wheelbase 2375mm
Overall length 3914mm
Overall width 1780mm

RANGE 330 miles
TANK CAPACITY 50 litres (11.0 gallons)

Made and sold by Rover Group Ltd, Bickenhill Lane, Bickenhill, Birmingham B37 7HQ. Tel: 0121 782 8000

The figures were taken at the Millbrook Proving Ground with the odometer reading 3500 miles. AUTOCAR test results are protected by world copyright and may not be reproduced without the editor's written permission.

SPECIFICATION

ENGINE
Layout	4 cylinders in line, 1796cc
Max power	118bhp at 5500rpm
Max torque	122lb ft at 3000rpm
Specific output	66bhp per litre
Power to weight	110bhp per tonne
Torque to weight	114lb ft per tonne
Installation	transverse, mid, rear-wheel drive
Construction	aluminium alloy head and block
Bore/stroke	80/89mm
Valve gear	4 valves per cylinder, dohc
Compression ratio	10.5:1
Ignition and fuel	MEMS electronic ignition, multi-point fuel injection

GEARBOX
Type 5-speed manual
Ratios/mph per 1000rpm
1st 3.17/5.3 2nd 1.84/9.2 3rd 1.31/12.9
4th 1.03/16.4 5th 0.77/22.1 Final drive ratio 3.94

SUSPENSION
Front double wishbones, Hydragas springs, anti-roll bar **Rear** double wishbones, Hydragas springs, anti-roll bar

STEERING
Type rack and pinion, optional speed-sensitive electric power assistance (fitted on test car)
Turns lock to lock 3.1 (3.4 without EPAS)

BRAKES
Front 240mm (9.0in) ventilated discs
Rear 240mm (9.0in) discs **Anti-lock** optional

WHEELS AND TYRES
Wheel size 6Jx15in **Made of** cast alloy
Tyres 185/55 VR15 (f), 205/50 VR15 (r)
Goodyear Eagle Touring **Spare** space saver

Mobil — AUTOCAR road tests are conducted using Mobil Unleaded or Diesel Plus containing additives to help keep engines cleaner

PERFORMANCE

MAXIMUM SPEEDS
5th gear 123mph/5560rpm 4th 115/7000
3rd 90/7000 2nd 64/7000 1st 37/7000

ACCELERATION FROM REST
True mph	sec	speedo mph
30	2.8	34
40	4.3	45
50	6.2	55
60	8.7	66
70	11.5	77
80	15.4	87
90	20.3	98
100	27.0	108
110	39.2	119

Standing quarter mile 16.6sec/82mph
Standing kilometre 30.2sec/103mph
30-70mph through the gears 8.7sec

ACCELERATION IN GEARS
mph	5th	4th	3rd	2nd
10-30	–	–	5.8	3.8
20-40	11.9	7.8	5.7	3.7
30-50	11.6	7.4	5.6	3.9
40-60	11.5	7.3	5.7	4.5
50-70	12.1	7.9	6.0	–
60-80	13.4	8.5	6.9	–
70-90	18.4	9.3	9.0	–
80-100	23.5	11.8	–	–

WEIGHT
Kerb (including half tank of fuel)	1073kg
Distribution front/rear	45/55 per cent
Gross vehicle weight	1320kg
Max payload	260kg
Max towing weight braked	n/a
Max towing weight unbraked	n/a

WHAT IT COSTS
List price	£15,995
Total as tested	£17,540

EQUIPMENT
(**bold type** denotes options fitted to test car)
Automatic gearbox	–
Metallic paint	£230
Driver's airbag	●
Passenger's airbag	**£345**
Seatbelt pre-tensioners	●
Electrically adjustable mirrors	●
Auto reverse radio/cassette player	●
Electric aerial	●
Power steering (electric)	**£550**
Alloy wheels	●
Adjustable steering column	–
Electric windows	●
Remote central locking	●
Tilt adjustable driver's seat	●
Anti-lock brakes	**£650**
Front foglights	£188 (DO)
Anti-theft system	●

● standard – not available DO dealer option

INSURANCE GROUP 12

WARRANTY
12 months/unlimited mileage, 6 years anti-corrosion, 3 years cosmetic paintwork, 12 months recovery

SERVICING
Initial 3000 miles, 0.5 hours, free
Interim/major 12,000 miles, 1.5 hours, n/a parts

PARTS PRICES
Oil filter	£7.32
Air filter	£12.22
Brake pads front/rear	£58.00/96.35
Exhaust (excluding cat)	£498.20
Door mirror glass	£18.80
Tyre (each, typical)	£129 (f), £160 (r)
Windscreen	£117.50
Headlamp glass	£38.78
Front wing	£82.25
Rear bumper	£141.00

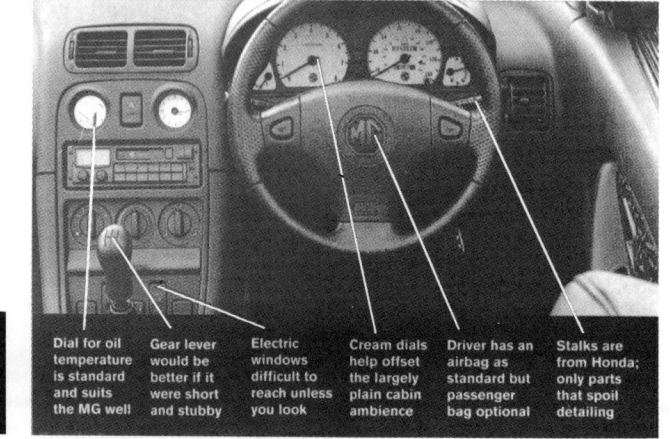

Interior callouts:
- Dial for oil temperature is standard and suits the MG well
- Gear lever would be better if it were short and stubby
- Electric windows difficult to reach unless you look
- Cream dials help offset the largely plain cabin ambience
- Driver has an airbag as standard but passenger bag optional
- Stalks are from Honda; only parts that spoil detailing

28

an engine behind the seats has ever felt this safe to drive fast.

But while we applaud this attention to safety, we lament the monopoly it holds over the MGF's handling characteristics. It provides no opportunity for altering your line with the throttle, no alternative to the steady understeer that arrives when you press hard through a corner. You feel you have no more than a front row seat at a show which, while beautifully produced, is to be appreciated more for the way it has been directed than the excitement it contains. By comparison the MX-5 gives you the lead role in an all-action blockbuster. The MGF's optional electric power steering is pleasingly weighted but lacks feel and has too many turns across a poor lock.

Ride
★★★★ Another deeply significant clue to the MGF's character: it rides superbly. Scuttle shake is minimal while it boasts a prodigious appetite for mid-corner bumps. Tackle any country road in an MX-5 followed by the MG and you'd swear someone had resurfaced it as you swapped seats. Half of this is the legacy of its mid mounted engine, a naturally ride-enhancing configuration thanks to the much softer front springs it allows. The other half is the Hydragas springing which, in the MG, has provided all the compliance we have come to expect from the system yet with none of the bounce. It is almost impossible to severely upset its suspension as even deep urban pot-holes are sponged away with little more than a small shudder. Only transverse ridges on country roads are felt particularly, a characteristic that's more apparent under braking. Even so, you'd be hard pressed to call it a serious fault.

LOVES
Fantastic feel through the brake pedal; oil temperature gauge; nicely detailed filler cap; properly designed and shaped boot; the looks it gets everywhere it goes

HATES
Terrible, fiddly hood bag; lack of on-board stowage space; plastic rimmed steering wheel feels cheap to the touch; gear lever should be two inches shorter

Handling characteristics are dominated by fine grip and poise. It's so safe but insufficiently involving for some

Brakes
★★★★ There's precious little to fault here. We spent an entire day lapping the Croft racetrack for next week's Best Driver's Car feature and not once did the four discs fail to slow the MG with anything other than masterful authority.

We would, however, prefer to have seen anti-lock brakes as standard on this car.

At the wheel
★★ The driving position is too high and restricted by its unadjustable steering wheel. Though the coloured instruments work well, the Honda-derived column stalks are nasty and fiddly to operate. Such a range of faults are a shame in such an otherwise complete car. The excellent visibility and competent ventilation system do little to make amends.

Accomodation and Comfort
★★★★ Unlike the MX-5, the MGF is a practical daily driver for someone without a family. The boot is big enough for holiday luggage as well as the weekly shop while the cabin boasts significantly more leg room, elbow room and hood-up head room than the MX-5.

The roof mechanism owes more than a little to the Mazda system, allowing you to snap undone the two retaining clips on the windscreen and throw the roof over your head. If you're quick and don't bother to unzip the rear screen (Rover says it's not strictly necessary), you can lower the hood in rather less than 5secs.

Noise
★★★★ Hood down, the MGF is extremely refined, proving resistant to wind buffeting at all speeds you're likely to reach in the UK. With the roof in place, it could almost be a coupe, so resistant is it to the intrusions of wind and road noise.

Build quality and safety
★★★★ The MGF feels hugely strong and beautifully built. Paint quality and panel fit exceed the impressive standards of the MX-5 and the sense of body integrity is perhaps unique among small open sportscars.

Safety features include a standard driver's airbag, seat belt pretensioners and side intrusion beams though a passenger airbag and, a little disappointingly, anti-lock brakes remain an optional extra.

Equipment and value
★★★ As you'd expect, the MG comes with alloy wheels, an alarm, central locking and electric windows. If you want power steering, anti-lock brakes or a passenger air-bag, they'll add £1545 to the list price.

On paper, the MG's price is at least competitive with its rivals while many might consider its engineering, construction and the magic of its badge to be well worth a few pounds more.

THE AUTOCAR VERDICT
★★★★ The MGF is going to sell like lottery tickets on a roll-over week. What's more, it will deserve to. It would have been so easy for Rover to have put an MG badge on the nose of a mediocre car and once more rely on the marque's image to do the selling. This has not happened. Rover has instead created what is, in all probability, the world's most complete and affordable open two-seater. From traditional MG fans to those wanting something more stylish than the chopped-about hatchbacks that pass for convertibles these days, the MGF should prove a blessing. It is an all-British car of which we can be unusually proud.

The only people who will feel left out are those who had hoped for a genuinely thrilling driving experience. For all its talents, it is the one service the MGF manifestly fails to provide. Many see this as the first priority of an open two-seater sportscar and some will rule the MG out on those grounds alone. And while they would be missing out on a truly great new car, we, enthusiasts all, understand how they feel.

DRIVING THE NEW CARS

MGF

It's been a long time coming – but after all the broken promises, we've finally got the real thing, says Ian Adcock

Remember the glorious-looking EXE which stunned the Frankfurt Motor Show back in the 'eighties? A mid-engined supercar that could have conquered the world. But it didn't and remains to this day a styling concept. At least EXE made it part of the way. But how many other drawings, concepts, studies, reports were produced, discussed and discarded? We'll probably never know. All we do know for certain is that Britain's once-independent carmaker – variously called BLMC, BLMH, Austin-Rover and finally Rover during those times – abandoned its position as the pre-eminent manufacturer of affordable sports cars and handed it over lock, stock and barrel to the Japanese – and Mazda in particular.

Admittedly, in those dark days the management had other concerns on its mind, like survival. The money wasn't there in the kitty to spend on niche products like sports cars.

As Rover's confidence (and profits) grew, so did rumours that it would produce a sports car, that the MG name – once relegated to decalled Metros, Maestros and Montegos – would again take its rightful place on showroom floors as an independent marque.

Arrival of the MGR V8

We waited and waited and nothing came. Then, a couple of years ago, the MGR V8 appeared: a limited edition run of Rover V8-powered two-seaters based on the antiquated "B" bodyshell. The Japanese loved it – to the extent of buying 1 200 of the 2 000 built – and so did I. Apart from its cost, it reminded me of what MG motoring was all about: it was quick – but not *too* quick – enough to have slidey-out tail fun without needing F1 reflexes to catch it. However, for twenty-somethings it was just old-fashioned.

But, and this is the important part, it revived interest in the MG marque. It heralded the Octagon's return. No longer was it a question of "if", but "when".

It's been a long six months since the MGF was first seen at Geneva earlier this year, where it wowed the press, punters and rival manufacturers alike. I can tell you now that every second has been worth it.

Normally, you wait 'til the end of an article like this to find out if the car's any good. But I won't do that to you today: the good news is, the wonderful news is, that MG is back with a car – give or take a few niggles that can easily be cured – equal to, and better in many cases, than the MX5, Barchetta, Elan, TVR, Alfa Spider *et al*. And, let's face it, it had to be that good to stand a chance in the market place.

Britain's dry, dusty drought of a summer had gone the week before I was due to drive the F for the first time, replaced by the remnants of a hurricane that had sidled across the Atlantic from the Caribbean and was lashing the UK with howling winds and teeming rain. Just my luck.

British Racing Green

I had programmed myself for a couple of days of hood-up soggy sports car motoring, despite the blue skies and sun that warmed the British Midlands as I drove to collect the test car. And I could hardly believe my luck when it turned out to be the British racing green

The MGF's success comes with any view that includes the front. The black "egg-crate" mesh of the grille conjures up images of bygone MGs.

The wait for the new MGF has been worth every second. There's not a wasted line or gram too much metal in its form (main pic). There's plenty of legroom (left), even if the cockpit feels narrow. The 1,8-litre K-series sits behind the driver and passenger, but it's not that accessible (right).

example that was parked at Rover's Gaydon test facility and not the Volcano (metallic orange), or the Amaranth (purple), examples that flanked it. The omens were looking good for the day.

Seen away from the lights and crowds of Geneva's crowded Salon you're immediately struck as to how compact the MGF is, how tightly formed between its wheel-at-each-corner stance. There isn't a gram of wasted metal, its tight packaging has been moulded around the minimal dimensions demanded to house an engine, two tall adults and some luggage.

Unlike some roadsters, the F looks good with the hood up or down. With the hood in place it takes on a far more wedge-shaped appearance, the rise from the headlights to boot lip more discernible than when the hood is stowed away.

Despite the boldly-shaped rear lights, it's the back view that is the most disappointing. The panel between the lights is bland, broken only by a single MG Octagon badge – the addition of a high-mounted stop-light might alleviate that. But it's the view beneath the bumper that really lets the back down. Twin vents give those following an excellent view of the silvered transverse silencer box, which also extends marginally below the rear lip. Lotus had exactly the same problem with the Elan and got round it by using black gauze between the vents to mask the silencer. Rover should do the same and also paint the silencer black or somehow shroud it. In fact, the F displayed at Frankfurt a few days later had a black silencer box...

Egg-crate mesh

The F's success, though, comes with any view that includes the front. The black "egg-crate" mesh of the grille, and

31

the manner in which the bonnet badge is mounted conjures images of bygone MGs – especially the B – without parodying them. Match that to the bold headlights and you have a car that is almost smiling at you when it comes down the road.

During the week-end, it was variously described as "dinky", "cute" and even "sweet" and, looking back, I think it's because the MGF looks like fun, inviting to drive.

There are also some nice styling touches – like the racing filler cap and a couple of stainless steel strips on the rear deck lid, complete with Allen key fixings. Nor is the car festooned with MG badges which would have been the easy and crass thing to do. The "F" doesn't need to shout "MG" – it *is* an MG. There are but two – one front and one rear – updated versions of the original cream and rusty red Octagon badges that adorned my MGA of 20 years ago (oh, how I regret its sale).

Hell, there's only one way to drive a convertible when the sun's up and that's with the hood down. After unzipping the rear plastic panel and laying it flat on what doubles as a small parcel shelf when the hood's in place, it's a simple task of unclipping a pair of over-centre catches and folding the (Pininfarina-designed) hood back. The fiddly bit comes when you try and fix the tonneau. Trying to thread the thin wire beading into its location to prevent ballooning at speed is tiresome and requires patience, and unless done exactly right, it soon comes adrift.

The rear aspect (top) is the unhappiest, with a large bland expanse between the taillights and the clearly visible silencer box. The front compartment (centre) is untidy, with little storage space. The boot (above) is deep, and adequate for soft bags.

Stern test

The hood itself is a single-layer affair which was given a stern test 24 hours later when that dying hurricane I mentioned earlier did a U-turn and swept back across the UK with lashings of rain. Apart from a minor whistle, where it didn't seat fully against the passenger's window, it was weather-proof and didn't balloon too much at speed. Apart from that, and the fact that just like my A there was the requisite drip of rain onto the seat as the doors were opened, the hood was rain- and wind-proof. The only driving problem was the broad B-posts, which make pulling out of acute angled junctions difficult, as it is in most rag-tops.

The interior is as intimate as the rest of the car is compact; driver and passenger sit almost rubbing shoulders, and there's plenty of legroom even for lanky 180+ cm poles like me. The car wraps itself round you, with the high door line and rear decks combining with the low seating position to emphasize that you're sitting in the car and are one with it.

It's a good driving position, straight-legged with your right foot naturally falling to heel 'n toeing position on the drilled throttle pedal. I could have done with the chunky-rimmed steering wheel being a couple of centimetres higher, but even so it never once detracted from the driving pleasure.

In your line of vision there's a simple instrument display, in traditional red figures on a cream background that echoes the Octagon badge. The minor controls you'll recognise from Rover's parts bin, and they're none the worse for that.

Twist the key and Rover's new 1,8-litre 88 kW engine springs into life with anticipation, blip the throttle and the needle swings easily round the tacho accompanied by an eager engine note just behind your ears. The fun factor is growing by the second.

Electrically assisted steering

The first few kilometres are taken gently as I play myself into the car, feeling my way round the five-speed 'box and the novel (optional in the UK), electric power-assisted steering. All the controls are light, but not sloppy; the pedal efforts are well-balanced, the stalks and switches operate with a satisfying "click" and the power-steering takes away the effort but none of the feedback while the engine responds immediately to the throttle.

There was only one niggle that persisted and that was with third gear. Hurried up-or down-changes – especially third to fifth – often resulted in a 'box full of neutrals as the linkage balked, character building if you're rushing down into a tightening bend.

As the roads open up and kilometres slip by, confidence gains and speed builds up. The Hydragas suspension absorbs all the ripples and pot-holes but simultaneously controls body movement so there's no pitch or wallow. Only on the roughest of surfaces does it get upset and feed through to the driver and passenger, the windscreen frame betraying the occasional shimmer.

A snaking series of bends unfurling themselves towards the horizon are attacked with gusto in fourth gear, with 140-150 km/h on the speedo, and succumb to a series of fluid steering inputs. The electric power steering is uncanny in the way it remains evenly weighted throughout the lock, there's little or no kickback, yet all the pertinent information is there at your fingertips. Neither is there any body roll to upset your chosen course, or to caution you to back off. The MGF does – to use that over-quoted phrase – seem to corner on rails.

Its overall attitude is neutral, shading perhaps to mild understeer at very high speeds. It reminds you of Lotus' standard-setting front-wheel drive Elan in the manner in which a small-engined, well-balanced car can eat up the kilometres with far less effort than a bigger car with more power.

The day was spent in convoy with a colleague in a Honda NSX Targa F-matic (not the best of the breed, it has to be conceded), and this man, no mean driver himself, was often hard at work maintaining station with the F, only closing or opening large gaps when superior horsepower came to the fore.

On the road, the MGF springs to life with its friendly "face" seeming to smile at you. Outright acceleration isn't blistering, but is lively enough. Ride control is excellent and very comfortable.

K-series motor

Rover's latest incarnation of the K-series is superb. Maximum torque – a healthy 165 N.m at 3 000 r/min – and 88 kW of power at 5 500 r/min are well below the 8 000 r/min red-line, but so flexible is the engine and so intoxicating its note that it would be a sad soul indeed who never used that limit just for the sheer hell of it.

The MGF is an invigorating, car to drive. In a year in which we've been seduced by Fiat's Barchetta and Coupé – OK a bigger car, but still a sportster – and succumbed to the rakish Alfa GTV and Spider, Rover had a daunting task on its hands with a new MG. Here was a company that has survived mainly by co-operating with Honda for its mainstream products, about to go out and develop from the ground up a new, mid-engined car bearing a legendary badge. Many were the cynics who said they didn't possess the depth of engineering talent to succeed. Well, they were wrong.

In the MGF, we have a car that not only looks fun to drive but *is* fun to drive. It possesses huge reserves of talent and character in the way it rides, corners, handles, performs. There are, undoubtedly, quicker straight-line cars, but sports cars are for the twisty bits, scything and arcing through swooping bends in the open air, and at this the F excels. At times 88 kW might feel insufficient, though for most it will suffice, but with the 107 kW Variable Valve Control version due shortly, that lack of muscle will soon be addressed.

But the over-riding impression of Britain's best two-seater to date is not of an MG built by Rover, but of an MG conceived and built by MG.

The Octagon is back. ●

MGF TECHNICAL SPECIFICATION

	1,8i	**1,8iVVC**
Engine:		
Cylinders	four	four
Fuel supply	multi-point fuel injection	
Bore/stroke (mm)	80/89	80/89
Cubic capacity (cm^3)	1 796	1 796
Compression ratio	10,5:1	10,5:1
Valve gear	d-o-h-c	d-o-h-c
Ignition	electronic	electronic
Engine output:		
Max power (kW/r/min)	88/5 500	107/7 000
Max torque (N.m/r/min)	166/3 000	174/4 500
Wheels and tyres:		
Wheels (front/rear)	6Jx15 inch	6Jx15 inch
Tyres (front/rear)	185/55 VR15/205/50 VR15	
Brakes:		
Front/rear	disc/disc	disc/disc
Steering:		
Type	rack and pinion, electric power-assisted	
Suspension:		
Front	independent, double wishbone, Hydragas springs, anti-roll bar	
Rear	independent, double wishbone, Hydragas springs, anti-roll bar	
Performance figures (manufacturers' claims):		
0-96 km/h (secs)	8,5	7,0
Maximum speed (km/h)	193	209

MGF 1.8i

BY GEORG KACHER

PHOTOGRAPHY BY MARTYN GODDARD

Long on ability, short on character.

Gaydon, England—

Fifteen years after the demise of the MGB and seventy years after the first two-seater droptop was completed by Morris Garages (MG), the Rover Group has launched a new roadster fitted with that prestigious octagonal badge. It's called the MGF, it's mid-engined, and it's a fine automobile. The only element that may spark discussions both among MG addicts and newcomers to the marque is the design. The interior in particular is a disappointment. Crude materials, drab colors, and a truly unimaginative dashboard ruin the driver appeal even before the engine has had a chance to fire. Unfortunately, the exterior doesn't qualify for the Museum of Modern Art, either. The proportions are quite well balanced, but from some angles the MGF looks rather tall and bloated. If it wasn't for the familiar grille, one could easily mistake this very British car for a middle-of-the-road product made in Japan.

Being a mere 154 inches long, this two-seater is about as tight-fitting as a leotard that has been left in the dryer too long. Tall drivers could definitely do with more head and leg room (not to mention an adjustable steering wheel), mechanics could do with easier access to the secluded engine bay, and travelers used to full-size suitcases could do with a deeper luggage compartment. The front-end cubicle is off limits for cargo because filling the crumple zone would affect the crash performance. Instead, this space is used by the spare wheel, the brake servo, and the battery.

The manual roof is a Pininfarina design. It looks good with the canvas top up or down, but the metal hood mechanism is a threat to the temples of driver and passenger; the rear window is of the plexiglass, scratch-prone, unheated kind; and the fit of the tonneau cover (which was given preference over a flush-fitting lid) reminds you of an adult's glove on a child's hand. Opening the roof would be even easier if you didn't have to unzip the rear window prior to the operation. Extra money buys a light yet sturdy hard top that quickly converts the roadster into an equally pretty coupe. Designed by Geoff Upex and Gordon Sked, who recently left Rover to open his own consulting business, the all-steel body is supplied by Motor Panels in Coventry. From there it is shipped to Cowley for final assembly.

Unlike other roofless beauties, whose bodies flex like Claudia Schiffer on the catwalk, the MGF's torsional stiffness leaves little to be desired. Among the chief reinforcement measures are two strong transverse box sections as well as a pair of unusually wide and tall door sills that have inner, outer, and middle walls. There is no roll bar, but the windshield frame has been beefed up for added stiffness and for the unlikely event of the car landing upside down. Thanks to the impeccable structural integrity, familiar vices like ill-fitting doors, cowl shake, and firewall judder are absent.

It may be misleading to claim that the MGF is based on the little Rover 100, but there is no denying that it shares the front and rear subframes as well as the Hydragas suspension with its hatchback sibling. Conceived by Alex Moulton, the Hydragas system uses gas- and fluid-filled spring and damper units interlinked front to rear. In the past it had not been considered fit for a sports car, but the project director, Nick Fell, and the chief engineer, Brian Griffith, decided to have another go at it because they were intrigued by the light and compact assembly. The Hydragas spheres are supported by double wishbones all around, by anti-roll bars at each end, and by conventional shock absorbers. The rack-and-pinion steering is power-operated, ABS is optional for the four disc brakes, and the stylish six-spoke aluminum wheels are shod with unequal-size tires (185/55VR-15 in the front, 205/50VR-15 in the rear).

The MGF is available with a choice of two 1.8-liter four-cylinder engines. The top-of-the-line version relies on a 145-bhp powerplant with variable valve timing. We drove the lesser 120-bhp edition, which sells in England for the equivalent of $24,800. Tipping the scales at 2337 pounds (that's only 44 pounds more than a Mazda Miata), the topless Brit will accelerate in 8.5 seconds from 0 to 60 mph, reach a maximum speed of 120 mph, and average 34 mpg in the city. The 145-bhp model is 1.5 seconds quicker off the mark and 10 mph faster overall, but it is also $3100 more expensive.

Dynamically, even the lesser MGF is a fine piece. Its single most outstanding feature is the brakes that decelerate the roadster with aplomb. Modulation is a delight, the pedal pressure barely increases under stress, and the stopping power reminds you of a roller coaster at the end of its ride. This car opens a new dimension in late braking, and the expertly tuned ABS makes sure that it's safe to drop the anchors even in the middle of a fast downhill bend. The steering is less convincing. Our test car was fitted with the optional speed-sensitive

MGF 1.8i

steering that employs a computer-controlled electric motor to provide its servo effect. Light and well damped, this system is sadly neither quite quick enough nor sufficiently communicative.

The sixteen-valve, 120-bhp, 1.8-liter four barely qualifies as a sports car engine. Its fortes include strong midrange punch, eager throttle response, and good fuel economy. Redlined at 6000 rpm, the unit's willingness to rev is hardly sensational, and the noises emitted by the twin tailpipes are more in line with a family sedan than with an open-air two-seater. Although the car by no means feels underpowered, its true potential is marred to an extent by the excessively tall gearing. Second takes you to 70 mph, third is good for over 100 mph, and the overdrive fifth is so long that it reaches halfway around the Oxford ring road. In sharp contrast to the transmission it is mated to, the pleasantly light clutch operates with laudable swiftness and inspiration.

Considering its humble origin, the Hydragas suspension does an excellent job. Its main strengths are total composure and sure-footedness. This car is so well balanced that you could drive it over a Bosnian minefield without spilling the coffee clutched by the fold-down dual cup holders. With the exception of very occasional lateral irritations, the ride comfort is in a league of its own, too, and this achievement counts double because it doesn't entail penalties like undue body motions or lackluster roadholding. Quite the contrary: The MGF hangs on like a leech and grips like an eagle, so there is no prize for guessing that its tires come from Goodyear.

The directional stability of a mid-engined sports car typically equals that of yours truly after four large whiskies, but the new MG makes an exception to this rule. It tracks with commendable accuracy, and it won't lose its equilibrium even if you back off in the middle of a corner tackled at the limit. Although the handling remains neutral for nine out of ten steps up the g-force ladder, the rear end will eventually swing round and display the oversteer that is inherent to the breed. With more oomph on tap, you could probably hold the tail out there almost all the way through the curve, but the 120 horses rarely permit more than a brief slide followed by an even briefer counterswing, which can be regarded as a gesture, not a threat. The steering mercifully refrains from loading up during such an exercise, but it isn't all that keen to get involved, either.

The MGF lacks the charismatic design and the raw power of the traditional RV8, yet it turns out to be much more refined, much more manageable, and, ultimately, a much better automobile. Having said that, the mid-engined roadster is too cramped and probably too expensive to stand a chance of being sold in the United States. Its replacement, however, and the Rover Group's upcoming all-new front-engined, rear-wheel-drive convertible that may end up using the Austin-Healey name, have already applied for a visa and a green card. Although the debut of these models is still about five years away, it never hurts to start saving early.

MGF 1.8i
Mid-engine, rear-wheel-drive roadster
2-passenger, 2-door steel body
Base price (in England) $24,800/price as tested (in England) $26,350

POWERTRAIN:
16-valve DOHC 4-in-line, 110 cu in (1796 cc)
Power DIN 120 bhp @ 5500 rpm
Torque DIN 122 lb-ft @ 3000 rpm
5-speed manual transmission

CHASSIS:
Independent front and rear suspension
Variable-power-assisted rack-and-pinion steering
Vented front disc, rear disc brakes
Anti-lock system
185/55VR-15 front, 205/50VR-15 rear Goodyear Eagle GT Touring tires

MEASUREMENTS:
Wheelbase 93.5 in
Length x width x height 154.0 x 70.1 x 49.6 in
Curb weight 2337 lb

PERFORMANCE (manufacturer's data):
0–60 mph in 8.5 sec
Top speed 120 mph
European city driving 34 mpg

COMPETITORS:
Alfa Romeo Spider
Honda Civic del Sol
Mazda MX-5 Miata

Oh, the joys of being behind the wheel of an MG roadster. Seeing that octagonal badge again is good for the soul. Passenger's air bag is optional.

UPFRONT

MGF:
New Duds for an Old-Fashioned Revival

From the 1920s, when Cecil Kimber started making hot rods based on humble Morris sedans, MG (for Morris Garages) has stood for affordable cars with a sporting flavor. Rarely innovative, inexpensive MGs had an easy charm that beguiled generations of young drivers on both sides of the Atlantic.

Until recently, MG's fame rested mostly on the various T-series roadsters, the MGA, and the MGB and companion GT, the latter a pair of fairly brisk and pleasant sports cars that engendered most fond memories of the marque. Despite the cars' popularity, management (British Leyland, pieces of which became today's Rover) never afforded a successor. In one of the many tragedies of the British motor industry, the MG was allowed to become a parody of itself, with anemic engines and ridiculous rubber bumpers. When the end finally came in 1980, it did not come proudly.

It has taken more than 30 years for the company that became Rover to commit to building a completely new MG sports car. And even now, true to history, it is based on high-volume production parts. Its very existence, in fact, depends on a joint venture with Mayflower Group, whose Motor Panels subsidiary stamps the new MG's steel bodywork.

And yet, the MGF turns out to be considerably more sophisticated than the MGs of the past—more sophisticated than even the similarly priced Z3 roadster from Rover's new parent company, BMW. When BMW bought the Rover Group in 1994, these two concurrent sports cars were the only direct clash of model lines, and both were too far along the development track to cancel or modify.

While the Z3 is built on a familiar platform, the MGF has a completely new structure. Its engine lies amidships, and it has an upper and lower control-arm suspension with interconnected Hydragas springs. There's also electric power steering. The engine is a hitherto unseen 116-hp 1.8-liter version of Rover's well-regarded K-series 16-valve four-cylinder. It's also available with infinitely variable valve timing, which brings output to 141 hp at 7000 rpm—more than the 138 hp of the Z3's 1.9-liter four.

The gearbox is a Honda design, which raises the question of the Japanese company's involvement in the MGF. Until BMW's stunning takeover last year, Honda owned 20 percent of Rover. It's known that Honda had worked on a mid-engined CRX, and the MGF's styling does look vaguely Japanese. But Rover insists that the MGF has been developed without Honda's assistance.

The resulting design doesn't draw attention to the mid-engined layout. The short front hood is relatively high, and the seats are set well back in the cockpit, giving the MGF a spacious cabin without the footroom deficiencies normally associated with this type of layout. The high, cut-off tail even houses a reasonable trunk.

There are two disadvantages to this tidy arrangement. A practical one is that the engine is virtually inaccessible unless one unbolts the tray that holds the retracted fabric roof. The other is more subjective: the seating position is rather high, and the car looks less low-slung than a sports car should be.

That impression is confirmed by the driving experience. Handling, steering, and roadholding are all first-rate. The Hydragas suspension—gas springs, with fluid interconnections front to rear—provides a supple comfort that seems at odds with the sports-car ethic. But this car can romp around England's narrow lanes at tremendous speed without the slightest fuss. At the limit of grip it understeers, without any tail-happy antics.

Although it's quicker than a Miata, the base MGF with only 116 hp feels rather inadequate. The models with variable valve timing (VVC), and an extra 25 hp, are more exciting. Rover claims that, by swinging the tach needle to 7000 rpm, the MGF with VVC can reach 60 mph in 7.0 seconds and top out at 130 mph. Outwardly, the only way to identify this more powerful version is by the wheels—alloys with five spokes instead of six. The faster model also has more elaborate standard equipment, including anti-lock braking and leather trim, and its sticker is $3000 higher.

It seems a pity that there are no plans to sell the MGF in the United States. Rover admits the car was designed with Americans in mind—much of the hot-weather testing was done in Arizona—but its Land Rover Centers are the company's only presence in the States. And, of course, it would be impolitic to sell it through BMW.

—Ray Hutton

Manufacturer: Rover Group
Birmingham, England
Vehicle type: mid-engine, rear-wheel drive, 2-passenger, 2-door roadster
Price (U.K.): $24,500–$27,500
Engine type: DOHC 16-valve 4-in-line, aluminum block and head, Rover MEMS engine-control system with port fuel injection

Displacement .... 110 cu in, 1796cc
Power (SAE net) .... 116–141 hp @ 5500 rpm
Transmission 5-speed
Wheelbase .... 93.5 in
Length .... 154.1 in
Curb weight .... 2400 lb

37

Twin Test BMW Z3 1.8/MGF 1.8i

Will the MGF have things all its own way this year or will BMW's new Z3 steal the show? David Vivian samples both

BROTH

Sibling rivalry: The MGF and Z3 typify the 90s roadster revival. MGF lacks aggressive stance of its German rival, but has the edge on performance

ERS IN ARMS

Twin Test BMW Z3 1.8/MGF 1.8i

1.8-litre engine from the 318i generates 115bhp

> 'The Z3 is almost as handsome as James Bond himself'

Roadsters are back where they belong: on our minds, in our garages and at each other's throats. After all the sham and tokenism of the 80s – when enthusiasts had to look to hot hatchbacks for affordable thrills – cars with two seats, soft tops and lifetime grin warranties are surging defiantly onto Europe's roads in a new wave of hedgerow-skimming hedonism.

Thank the Mazda MX-5. It showed that the deliciously pure and simple pleasures of a light car with a sweet four-pot twin-cam in the nose driving lightly-rubbered rear wheels via a flick-wrist gearbox (a formula finessed by the original Lotus Elan) had been too long forgotten. Their resurrection – and the 'seat-of-the-pants' responses that came with them – caused a sensation.

That catalysed the revival. And with growth has come rich diversification and fierce competition. You can live your sporting life with engine and drive to the north, engine mid-ships driving the rear wheels or, naturally, stick with the classic front-engine, rear-drive layout. In truth, the sports car makers' instincts, here, are led almost entirely by engineering precedent. It would have been remarkable indeed if the Fiat Barchetta and Alfa Spider had turned out to be anything other than fwd. Not very purist but when your cars look and sound as good as these, there are compensations.

The new, and breathlessly-anticipated, BMW Z3 you see here – assembled from familiar-looking Munich-spun parts in Spartanburg, South Carolina, USA, and on sale over here in September – is no less predictable. From some angles it's almost as handsome as James Bond himself, who steered it through the movie *GoldenEye* with suave assurance last year in the marketing coup of the decade. But its running gear is essentially that of 3 Series Compact – an orthodox north-south arrangement – fitting the sports car brief nicely, if a little complacently.

You see, the Z3 gets the Compact's trailing arm rear suspension (itself a derivative of the previous-generation Three's) rather than the superior Z-axle of the 3 Series proper and the Z3's ill-fated forerunner, the Z1. It also gets, in standard (circa £20,000) form, the second weakest engine in BMW's armoury. Even so, the eight-valve 1.8-litre 'four' from the 318i delivers a respectable 115bhp at 5,500rpm and quite encouraging 123lb ft of torque at 3,900rpm.

The Z3 is reasonably svelte at

MGF cabin is sporty but lacks the finish of the BMW's. Seating position is too high for tall drivers

Only ancillaries to be seen under the MGF lid

'At £15,995 it looks nothing short of a bargain'

1,175kg but the resulting power/weight ratio of 98bhp/tonne seems disquietingly at odds with its provocatively long bonnet and those shark-gill side vents (nice retro touch, this, redolent of the 507), never mind the brutally macho wheel arches and the heavy-duty message sent out by the almost truck-tough 225/50 ZR 16-shod alloys. What a dainty rear end, though. Incongruously so, but, in this guise, more in keeping with the BMW's firepower.

It isn't only in the engine room that the MGF 1.8i – ironically, a much more feminine-looking machine – shows the Z3 up. But that's where we'll start. Its K-series motor has 1.8 litres, too, but with a twin-cam head and 16 valves pushes out 118bhp at 5,500rpm. Peak torque of 122lb ft is slightly down but developed at a more useful 3,000rpm. As well as being more compact than the BMW it's also some 115kg lighter and ends up with a much healthier 111bhp/tonne.

More telling than what the engine dishes out though, is where it sits: transversely behind the cabin driving the rear wheels. The easy thing for Rover to do would have been to make the MGF front drive a la Barchetta. But it went out of its way to do the job properly. If you're going to build a sports car by the latest rules, you don't start by putting the engine anywhere but in the middle. Get it right and there are benefits in response, grip and comfort.

That's why you'll experience fewer deja vu moments perusing the MG. Getting it right required more original thought. From the application of its double wishbone and Hydragas suspension to the design of its facia (admittedly not wonderful), the British car is more or less purpose-built. And, at £15,995, it looks nothing short of a bargain on paper. The Z3, on the other hand, is much closer to being a Compact wearing a muscular roadster body-suit. More auto-suggestion than form following function.

So a pretty convincing moral victory for the MGF, but does it translate into real-world supremacy? A BMW, after all, is a BMW. The Z3, despite its slightly confused gender, has more road presence than the soap-bar-with-attitude MG. More perceived prestige and class. Although Rover and its parent company BMW play on the same side these days, you'll never convince anyone that an MG can be as well made as a BMW. It isn't. Not quite. And while you've seen them before, the BMW's instruments are clearer and more reassuring than the MG's.

41

Twin Test BMW Z3 1.8/MGF 1.8i

BMW hood is a doddle to put up or down

MGF hood is fiddly to fit, but once in place it works well

While the Z3's tail is neatly sculpted, the MGF's podgy rear end hides its greatest attribute – the 118bhp K-series engine

Z3's boot far more practical than MGF's, but not a great deal bigger

It's just the beginning...
More powerful Z3s set to follow the 1.8 base model into the showrooms

There are some truly mouthwatering Z3s on the way. We've already seen BMW's vision of what a 911-eating M-Sport version might look like (see page 10) and, next spring, a version powered by the 328's cracking 193bhp 24-valve 2.8-litre straight six goes on sale.
Having driven the Z3 1.9, it's hard not to conclude that until those beefcake powerplants arrive, the car will be selling itself short. The 1.9 140bhp 16-valve engine comes out of the 318 ti Compact. Thus powered, the Z3 costs the equivalent of £21,800 in Germany, a modest enough price hike. But the boost in performance is hardly shattering and

the figures speak for themselves. The 0-60mph time drops from 9.8sec to 8.9sec (still not as good as the bog standard MGF) while the fifth gear 50-75mph time is cut from 14.5sec to 13.7sec. The 1.9 does its thing a little more effortlessly than the 1.8 but, that said, its broad powerband is also devoid of any real excitement; the tinny rasp from the exhaust is no substitute.
The extra muscle, subtle as it is, injects a little more balance into the chassis, but not enough to warrant the optional sport pack suspension BMW will be offering which comprises a limited slip differential and stiffer springs.

More powerful Z3s are on the way

KEY FACTS

Model	BMW Z3 1.8	BMW Z3 1.9
Engine/cylinders	Inline 4	Inline 4
Capacity ccm	1,796	1,895
Power bhp/rpm	115/5,500	140/6,000
Torque lb ft/rpm	124/3,900	133/4,300
0-60 mph secs	9.8	8.9
0-100 mph	30.2	25.5
50-75 mph (5th)	14.5	13.7
70-0 mph (cold/hot)	50.5/51.0	47.9/50.5
Max speed mph	124	127
Test figures mpg	31.4	29.5
Price (Estimated) £	20,000	22,000

Hoods at twenty paces? No contest. The MGF's is very good – taut, water-proof, a modest generator of wind noise – but the Z3's is a miracle of precision fit and user convenience. In the time it takes to fit the MG's absurdly fiddly tonneau cover, the Z3's entire hood would have been stowed, its driver's right foot reacquainted with the accelerator and a comfortable cruising speed attained.

But it wouldn't be long before the MGF zipped by. The squat British roadster has the legs of the Bimmer in nearly all departments. It sprints to 60mph in 8.7secs (9.8), 100mph in 26.6secs (30.2) and surges on to a top speed of 126mph (124). Only between 50 and 75mph in fifth does the Z3 exact any kind of revenge, covering the increment in 14.5secs against the MGF's 16.3secs. When it comes to low-down lugging power, though, the positions are once again reversed with the MGF disposing of the fourth gear 40-60mph haul in 9.8secs (10.3).

All right, the MGF isn't a quick car by Golf VR6 standards but it's tigerish compared with the Z3 which, subjectively, feels even slower than the figures suggest. The MG sounds more urgent, too, with a cultured, back-of-the-throat growl when it's trying. You can't fault the BMW's engine for smoothness and refinement – but it feels worryingly anaemic. To the extent that, under full acceleration, you find yourself leaning forward in the seat, urging it on. Just as well its short-throw gearchange is lightning-fast. Every fraction counts. The MG's

42

K-series engine encroaches into the boot space but there's still a reasonable amount of room

shift is snappy, too, but rather notchier and slower.

The BMW has a terrific chassis, it's as simple as that. Its steering is meaty, accurate and feelful. Poise is remarkable. It turns in, it hangs on and it inspires oodles of confidence. An MX-5 for grown-ups. With the weedy 1.8 doing the bidding, though, you get a nose-led cornering balance in which the throttle has little say. The abilities of the chassis are underexploited. Which is frustrating. As is the over-large steering wheel and the seats' lack of lateral support.

You sit higher in the MG but it feels handier and its seats hold you in place. They need to. They have to contain greater loadings and convey a wider range of moves to the base of your spine. The MGF simply has a larger palette of dynamic colours than the Z3. It grips harder but within its limits is a level of interplay between steering and throttle the BMW doesn't even hint at. Its combination of composure and communication at speed on a demanding road is breathtaking. Adjustability goes all the way from mild understeer to full-blooded oversteer, but the oversteer is 'real', not the gratuitous tail flicks an MX-5 driver can indulge in on any damp second-gear bend (the Z3 can't even manage that). Master it and the satisfaction is all the greater.

In the end, that's what sets the MGF apart. It's a serious sports car for serious drivers. We can't help feeling that the entry-level BMW Z3, for all its inherent goodness, is working to a different agenda.

Complete Car VERDICT — MGF captures roadster crown

1. MGF 1.8i £15,995 ★★★★☆

✓ The MGF isn't a family hatch in fancy dress but a purpose built sports car with its heart (and engine) in the right place. Sixteen-valve K-series unit has plenty of zip and character but better still is the superb chassis which just gets better the faster you go

✗ Unlike the Z3, the MGF actually offers a far more aggressive driving experience than its soap-bar looks would suggest. Seating position is too high for tall drivers and hood isn't as good as the Bimmer's. Facia design is original but something of a let down

2. BMW Z3 1.8 £20,000 (Est) ★★★☆☆

✓ It's a BMW and a good-looking one at that. Styling isn't as cohesive as MG's but has more presence on the road. The chassis is superb, as you would expect, and the steering perfectly weighted. Handling is crisp and confidence inspiring.

✗ Engine smooth and refined but feels limp-wristed given the Z3's weight. Interior holds few surprises – facia is neat and functional but too similar to a BMW saloon's. Steering wheel feels cumbersome and seats lack side support. Dainty rear doesn't match macho nose.

SPECIFICATIONS — MGF wins on performance

		BMW Z3 1.8	ROVER MGF 1.8i
ENGINE			
Capacity	ccm	1,796	1,796
Max power	bhp at rpm	115/5,500	118/5,500
Max torque	lb ft at rpm	124/3,900	122/3,000
KERB WEIGHT	kg	1,180	1,060
PERFORMANCE			
0-40mph		4.2	3.9
0-50mph		6.7	5.9
0-60mph		9.8	8.7
0-100mph		30.2	26.6
TOP SPEED	mph	124	126
40-60mph (4th)	secs	10.3	9.8
50-75mph (4th)		10.7	10.5
50-75mph (5th)		14.5	16.3
NOISE LEVELS			
Idle	dB(A)	50	47
Constant 30mph		62	65
Constant 50mph		68	68
Constant 70mph		72	75
Constant 100mph		81	84
BRAKING			
70-0 mph cold/hot	metres	50.5/51.0	51.9/52.6
FUEL CONSUMPTION			
Composite	mpg	37.1	42.1
Overall test		31.3	27.2
Best/worst test		44.0/22.8	46.2/20.3
EQUIPMENT			
Airbag (Driver)		•	•
Airbag (Passenger)		•	£345
Central locking		•	•
CD player		–	TBA
Electric seats		•	•
Hard-top		–	£1095
Leather upholstery		£1,020 (Est)	•
Metallic paint		£350 (Est)	£230
Seat height adjustment (Driver)		•	–

• Standard   – Not available

44

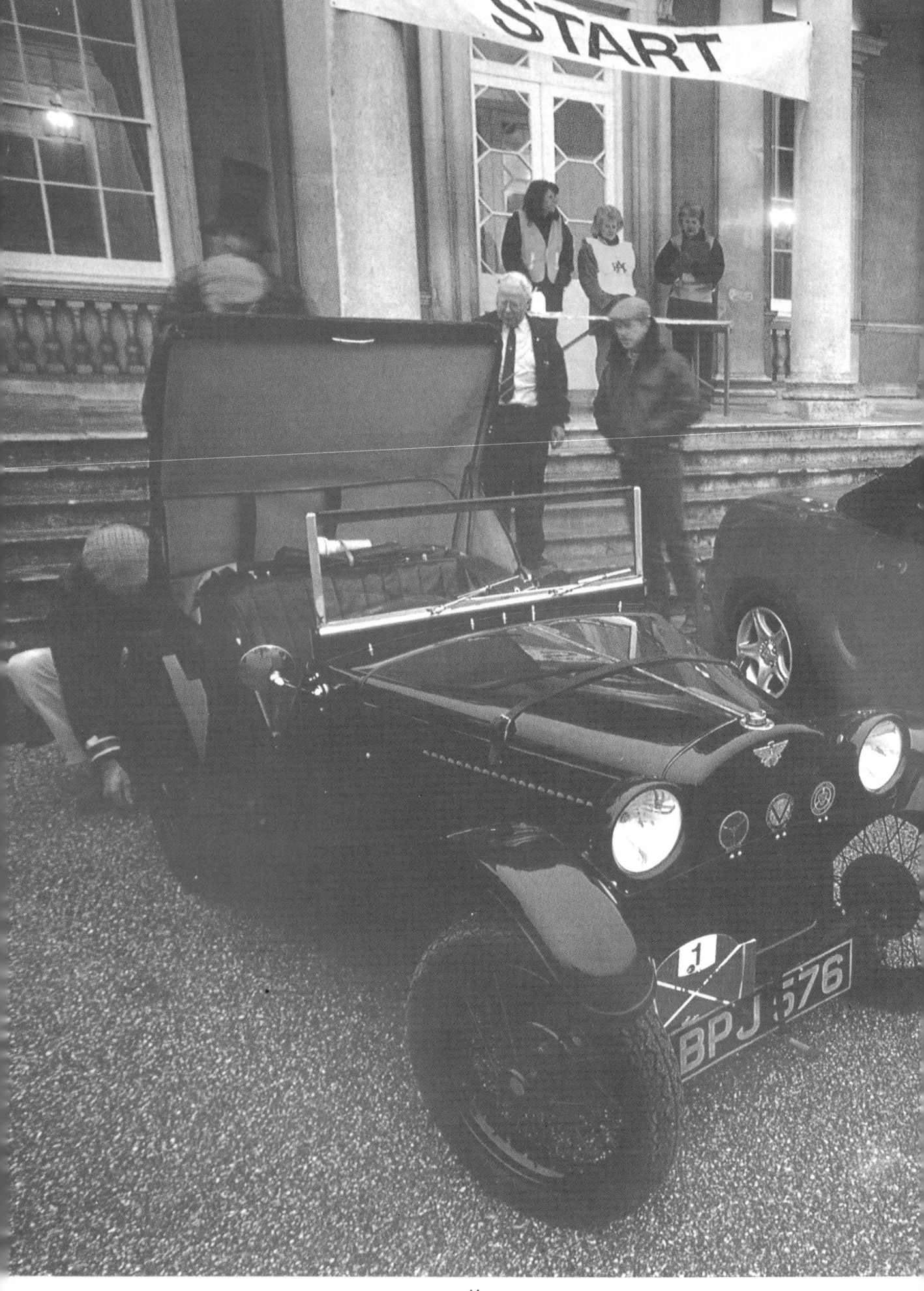

software

The **MGF** makes a big deal out of its **family history**: the MG **octagon** is displayed like a **valued heirloom**. So we took an **MGF VVC** on the high road to **Scotland** in company with some real **old warhorses** – retracing the route of the 1927 **London** to **Edinburgh** Trial along with some of the little MG's long-lost **descendants** to find out if it's the **black sheep** of the family or the **prodigal** great-grandson

Story: Tom Stewart

Photography: Philip Lee Harvey

CLAN GATHERING

Let's face it, if you had to follow 50 separate route instructions which haven't been updated since 1927, you'd get lost too. Here's a sample, taken straight from the Roadrunner Club's road book for their '96 London-Edinburgh Trial:

'L at Buckden near PO. Narrow by Hubberholmes. At second gate (Beckermonds) R for Hawes by Oughtershaw and Fleet Moss. At summit (1,852ft.), square R corner.'

Gibberish? Very nearly, but that's how they managed in the old days, y'know.

And that's how 60 classic cars (well, 58 plus our N-reg MGF VVC and a recent MG RV8) and their crews were expected to manage too, on a weekend recreation of the 1927 London to Edinburgh Trial; a gruelling and very competitive event in its day but a fine excuse for a weekend's moderately taxing fun today.

However, the taxing bit was mainly that the directions were taken directly from the '27 event's road book. And during over 400 miles you're bound to get lost. Not just a little bit, once or twice, but seriously and often.

For all I knew, other landmarks like 'ET Stead & Co's garage' or 'the Bacon Factory about 600yds N. of the second milestone from Carlisle' were levelled decades ago and their sites are now homes to microelectronics firms and TGI Fridays. So photographer Philip Harvey and I decided that, on our first London-Edinburgh Trial, we'd cheat.

In truth, cheat isn't quite the right word. We were determined to retrace the fun part of the '27 route just as accurately as everyone else – but with the help of a detailed Ordnance Survey road atlas and a pink highlighter pen I brazenly cribbed the trickier latter half of the road from Harrogate to Edinburgh straight from the organiser's own maps.

Frankly, we needed all the help we could get. The car should have been capable enough – you'd expect a 6,500-mile MGF VVC to be reliable – but our organisation wasn't quite as well-ordered. Even before the start I had inexplicably lost the all-important road book containing, not least, the directions to the start.

We were lucky. Having searched for it everywhere in London before sunrise and failed, we gambled on finding another trialist at South Mimms services (at least I could remember that much). Fortunately, the little red Midget we spotted at the pumps turned out to be driven by none other than the organiser, Ms Roadrunner herself, Kim Myers. Kim was a little better prepared than us (not difficult) and instantly provided us with a replacement book.

A '35 MG NB tourer parades past a line of young whippersnapper MGAs, MGBs and our MGF at Wrotham Park (above). Most MGB owners don't seem to find the new MG sexy, but this 'B' kept winking at our 'F' (right). Former F1 driver Philip Robinson (below, left) and friend Ivan Bailey give it the full Biggles 'chocks away' pose

Luxuriating in a sense of deep relief we breakfasted, then cheerily set off for Wrotham Park, the Trial's original starting venue in '27.

Technically, club rules state that, at the organiser's discretion, only pre-1980 classics are eligible. We'd been allowed to tag along, but upon arrival we were, rather embarrassingly, directed to line-up right next to the oldest car entered, Bill Higgins' 1934 two-seat Austin Seven special. Although the London-Edinburgh is now a non-competitive event, without any timed stages, it's still a serious journey in midwinter in a car like that.

By 9.30am Phil and I were on our way

software

spacious. There's a fair bit of wind noise from both the top of the screen and the hood, and the Hydragas springing is set on the firm side. Not bone-jarringly so, but you certainly know you're in a sportscar. No bad thing in my book.

I was slightly more concerned about remaining comfortable. Phil seemed happy enough but I was shifting to and fro, wishing there was some adjustment on the steering wheel. For me, it's set a couple of inches too low and can interfere with simultaneous steering and de-clutching. This less-than-perfect posture is exacerbated by the car's offset pedals. However, following some extended fidgeting, I finally found a position in which I remained content.

By lunchtime, and in spite of fog, we'd checked in at Harrogate's Majestic and, feeling just a little fatigued, spent a lazy afternoon drinking a variety of delicious ales in the Rat and Parrot before joining the others in the evening for aperitifs, dinner and more drink.

Day two, the route to Edinburgh, would be tougher. It took us up into the Yorkshire Dales and along the spine of the Pennines. Here the roads were either wet, muddy, icy, or all of the above. In addition, the fog was so thick that only a fish could breathe and the wind was cutting straight through 99 per cent of all clothing. Even if the F had been fitted with front fog lights we'd still have been navigating by touch.

We stopped for a chat with two fellow Roadrunners; Gloucesterman Alan Matthews and his dog-hatted co-driver in their cosmetically-something-short-of-concours '69 Herald convertible. We'd followed them for several miles over some treacherous roads and, frankly, were surprised they'd come this far, bearing in mind their vehicle and exuberant driving style. But they were unfazed. "Last year we had four foot of snow up here. That was fun!"

Still confused, we sealed ourselves back inside the MGF, checked the heater was still set to 'roast' and headed through the murk towards lunch at the Tan Hill Inn. Apart from being England's highest pub it is, significantly, where the late Ted Moult uttered those immortal words: "Fit the best – fit Everest!" It was also where many of the Trial's navigators were relishing their second party inside 12 hours.

Along the Pennine Way towards Penrith the fog lifted, which allowed me to exercise the F a little more energetically. In tight turns, either in town or in the hills, the car is a doddle to drive. The F's power-assisted steering works brilliantly and when the road opens up its speed-sensitive electronics add just the right amount of weight. Cornering at higher speeds, and with a few bumps thrown in, the firm suspension comes into its own and the F feels especially well-

Our Tom does his best to blend in with the masochistic eccentrics by dropping the hood of the MGF in freezing temperatures and donning the full Noddy-style wooly hat (left). But he is out-weirded by the canine-crested co-driver of Alan Matthews' '69 Triumph Herald Convertible (bottom left). They're mad, the lot of 'em

north. In common with most other London-Edinburgh runners we'd elected to use the A1 as it is today as far as Wetherby, and then turn off to the overnight stop at Harrogate, rather than uselessly attempt to navigate The Great North Road precisely as it was 69 years ago.

Unlike most others who, despite wearing up to 30 layers of clothing, are quite blatantly masochistic and proud of it, we kept our MGF's heater on and the cloth hood tightly closed. For more effective evaluation of the MGF's cabin characteristics, you understand.

Tucked inside at dual carriageway speeds, the F is neither particularly smooth, quiet or balanced and firmly planted. The Variable Valve Control engine contributes towards this F's *bona fide* sporting persona, too.

A few months earlier, a drive in the 118bhp 16v version had left me feeling entirely ambivalent. Quite simply, despite the mid-engined layout and all the hype, I didn't find the car remotely exciting. But the 143bhp VVC unit is significantly stronger and more responsive and, importantly, its exhaust note is much crisper.

For me, the VVC engine transforms the F from being a mere open two-seater into an exhilarating sports roadster – and it seems I'm not alone. Many MGF customers feel the same way and, as Rover have underestimated demand for the more powerful model, its price has recently been hiked to £2,400 above the cost of the basic 16v. But in my view the VVC is still worth the difference, especially when you consider its standard ABS braking, speed-sensitive power steering and half-leather seat trim. (The patterned cloth in our test 16v must have been chosen from the Furnishings for Oriental Restaurants catalogue.)

From Penrith the Trial continued up the Roman road to Carlisle, over the border towards Lockerbie and then took a right at Moffat to take us right out into the wilds again. It had been tempting to let the F really sing, but during daylight hours our duty was to stay in touch with the other, older cars on the run and capture their progress on film. And at this point may I say how refreshing it was to drive in the company of other enthusiasts; people who have both their eyes and their minds on the road and who know, for example, how to safely negotiate a roundabout at something above walking pace. I wouldn't deny that the speed limit wasn't occasionally infringed by some, but I'd rather drive with this band of eccentric Biggles lookalikes in their Midgets and MGBs than ninety percent of the other incapable zombies on the road.

With several more rolls of film in the can we found ourselves behind schedule by the time we reached Moffat. We'd lowered the roof earlier but by now it was dark, drizzling... and we were still behind Bill in his '34 Austin.

Fortunately, we had arguably the best driving road in the UK ahead of us; the A708 to St. Mary's Loch at Cappercleuch. The topography here is such that, in the dark, any other vehicles on the road, either ahead or behind, can be spotted miles away. To drive this deserted stretch with knowledge of what lies ahead must be huge fun but without that familiarity, and with mud-splattered headlamps, I had to rely on gut feel and the MGF's chuckable handling and reassuring brakes. I was driving at maybe six- or seven-tenths; but I doubt we'd have made any more time over this road's countless blind brows if we'd been in a Carrera 4 or NSX, for neither are as nimble as the F.

But, still trailing, we pressed on, (roof still down please note), towards Edinburgh. Somewhere on the A701 we overtook John and Jill Collier in their fixed-head '61 MGA and finally hit the city's ring road complex with plenty of time to make it to the hotel in time for several celebratory sharpeners before the dinner and disco. Not.

Within a mile of the Forth Bridges Hotel we got hopelessly lost. My temper could have been

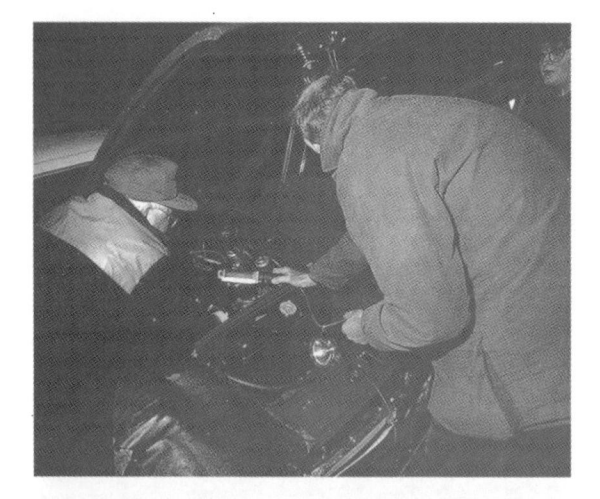

You've go to expect a few ailments among all the old cars on a long run like this. Luckily, the support crew were on hand, diagnosing advanced lumbago of the manifold in this case (top). The MGF shows off its sprightly young legs, passing a Triumph Herald-engined Burlington Special and blowing a big raspberry as it goes past (right)

48

software

persisted, so there was icing on pretty much everything, circuit included. Fortunately the sleet didn't freeze solid on the track so judicious lappery provided a whole lot more fun.

Again, the MGF proved what a well-sorted little sports car it is. Off-line, the sleet made the track virtually undriveable but on-line it was possible to get it slip-sliding at both ends with a commendable degree of predictability. Dial in too much lock and the F would gently understeer. With a little too much throttle, it would go loose at the rear. Too much of both and the Goodyear Touring tyres would lose grip all round – but nothing that couldn't be simply and safely brought under control. The ABS-equipped brakes also proved to have terrific feel on such a dodgy surface and kept the show on the road even when I left things a little too late. Ten out of ten to Rover in these departments.

Lower marks, though, in other areas. The fuel gauge needle drops like a stone below a third full and there's no low-fuel warning light; we ran out on the M6 going home (354miles on 11 gallons is 32.2mpg, including fifteen laps of Knockhill). The gearchange on this VVC sometimes balked a quick shift from second to third, and reverse was often difficult to find, graunching in protest. The hood catches didn't line up squarely and were a fiddle to secure. One mirror dropped out of position every time I closed the door. Rear three-quarter vision is very limited with either the soft top or optional hardtop in place. A couple of times it was reluctant to start and backfired after extended use of the starter motor once – obviously something amiss with the engine management.

And last, but by no means least, not one person I spoke to on the Trial (most of whom were MG owners) confessed a love for the MGF's looks. Some were indifferent and considered it 'OK' or 'quite pretty' but absolutely nobody said it was drop-dead gorgeous. Not even the MGB owner who had already put down a chunky deposit on one.

There was a lot of interest, however, in how much luggage the MGF could carry. Well, Phil and I took winter clothing for two days and nights and posh clothes for Saturday night, plus a camera bag, a lens bag, a film bag and a tripod. All of this fitted in the 210-litre boot and on the parcel shelf (with the hood up). You can also get a boot-mounted luggage rack.

After all that, it only remains to say that every one of the 60 entries completed the Trial's route. But I doubt any of us would have done so without one crucial piece of equipment – a highlighter pen. By the way, who nicked mine? □

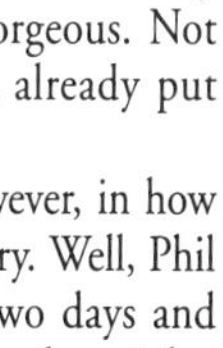

After the processional congestion of the open road, a thrash round the track at Knockhill allows the MGF to open up its lungs and have a good spit (above). Oh dear, looks like a touch of bronchitis for that MGB (top left). Rachel and Fred Larkin's Sanatogen-powered Moggy took to the track like an old whippet, though (bottom)

measured in millimetres and, dammit, we were thirsty. In addition, Phil was trying not to show it, but was fairly eager to get better acquainted with the enchanting Esther, who had navigated her father's '68 MGB GT from Holland...

It wouldn't be wise to recount the rest of the evening's events here, except to say that Bill Higgins, ably assisted by his son David, completed the Trial in their ancient Austin and on behalf of *Top Gear* I presented them with a small but well-deserved award for doing so.

Next morning we were all up bright(ish) and early for the icing on the cake; a blast around nearby Knockhill. The inclement weather

49

Lotus Elise v MX-5 v MGF

Champs
Elise

The stark Elise is the first real Lotus for a generation, but can its no-frills stance trounce the friendliness of the MX-5 and MGF?

Perhaps it's only now – six years on – that the UK sports car market is feeling the full effect of the ripples caused by Mazda when it dangled a speculative toe in the water with its MX-5.

Well, the little Japanese roadster proved that the world was ready for a '60s-style open-top that delivered '90s-style reliability and refinement. Throughout the decade, the MX-5 has quietly persuaded enthusiasts that it's okay for them to drive open-top two-seaters again.

Those early ripples have grown into a wave as other manufacturers scurried to produce affordable sports cars.

We already know about Rover's successful rejuvenation of the MG marque with its fabulous MGF. Demand for this car has been so great that low-mileage used examples are changing hands for more than they cost new.

Now it's the turn of Lotus to produce its first all-new car in six years, the Elise.

The Lotus name brings with it a bag of expectations: expensive, technically advanced, hand-built cars; elegant engineering solutions; low weight and high performance. All of these apply to the new Elise – except one. It has just gone on sale for £20,415, making it the cheapest Lotus for some years and pitching it into direct competition with the £18,795 Rover MGF VVC and the £18,010 Mazda MX-5 1.8i S.

Affordable? Certainly. Fun? Let's find out. ▷

LOTUS ELISE V MAZDA MX-5 V ROVER MGF

Rover MGF 1.8 VVC
£18,795
Hugely talented roadster that's every inch an MG

Mazda MX-5 1.8i S
£18,010
Japanese retro-sportster that began two-seater revival

Lotus Elise
£20,415
Keenly priced, no-compromise pocket supercar

Lotus Elise v MX-5 v MGF

PERFORMANCE
Lotus ●●●●●
Mazda ●●●
MG ●●●●

We would normally use the cars' respective engine and power outputs as broad guides to performance, but this trio is different, as we will see. All use 1.8-litre, twin-cam 16v engines driving the rear wheels; the Lotus and the MG, however, mount their power plants transversely behind the seats whereas the Mazda takes a traditional approach, with the engine mounted lengthways up-front.

The reason for the shared engine/transmission layout is that Lotus and MG use the same 1.8-litre K-series Rover engine, although this MGF VVC model goes one better by employing variable valve timing to enhance both engine power and torque (pulling power) across the rev range. The Mazda develops 128bhp and 118lb ft of torque, the MG 143bhp and 128lb ft, and the Lotus 118bhp and 122lb ft.

But these figures are deceptive, and the real clue to performance is weight. The Elise tips the scales at just 690kg thanks to an extruded aluminium spaceframe chassis that's epoxy-bonded. Yes, underneath its composite body, the Elise is glued together.

Now, consider that the metal-skinned MG and Mazda weigh 1070kg and 990kg respectively, and you can see why, on the road, neither can see the Elise for dust.

Check the specification table (p87) for the full low-down on performance figures, but the most important figures taken from there tell a heady story. Acceleration through the gears between 30-70mph – perhaps the most important real-world figure – takes the Mazda a respectable 9.8sec. The MG is appreciably quicker at 7.7sec, but the Elise blitzes this benchmark in just 6.1sec. To give that some perspective, that now-defunct small hatch supercar, the Ford Escort Cosworth, did it in 6.3sec.

It's the same story on the motorway. With the cars held in top gear, the Mazda takes 11.2sec to accelerate from 50 to 70mph and the MG 9.7sec. The Elise, shifting so little weight, does it in a mere 8.1sec.

From the bare figures, you might think the Elise has the best engine but, in fact, this accolade goes to the flexible MG. Its variable-valve-timed engine pulls brilliantly from low revs and has extra bite in the upper reaches of its rev range that the Elise's cannot match. The rorty MG also sounds better than the Elise, which emits a characterless blare.

There's nothing wrong with the Mazda's engine; it has plenty of urge and a pleasing rasp, but it lacks sparkle. Its gearshift which is the most precise of the three, though. The MG and the Elise use essentially the same 'box but the Lotus's shift feels ponderous and the lever looks as if it's borrowed from a tractor. Changes in the MGF are best made slowly, but the action is better damped than in the Elise.

Lotus Elise £20,415

Drilled pedals look handsome; no spare, but aerosol should get you home

'The Elise is a no-compromises sports car – and feels like it'

Minimal instruments and controls. No fuel gauge – an LCD readout tells you when the tank's empty

ROADSTERS GROUP TEST

HANDLING AND RIDE
Lotus ●●●●●
Mazda ●●●
MG ●●●●

No doubt about it, the Elise sets a new standard for driver enjoyment in this and, arguably, any other class of car. Again, it's the comparative lightness that gives it such agility. Suspension is by race car-like double wishbones all round; it also has narrower tyres on the front, wider ones behind. It has been set-up as a no compromises sports car – and it feels like it.

The ride is firm but never jars, and crests and dips are beautifully ironed out. There's little perceptible body roll in corners. In short, the Elise's chassis does everything so well you quickly forget how efficient it is. Only the brakes (made from aluminium – a first on a production road car) are disappointing. They're lighter than steel discs, of course, but their feel is no better, nor do they appear to shorten braking distances.

It's a delicate handler, the Elise. The unassisted steering delivers a constant stream of information about what the front wheels are doing, and over a bumpy road the wheel feels alive in your hands.

On a dry road, the Lotus generates massive grip and only the foolhardy would discover its limits. We took it to a test track and found the tail would happily step out under power, especially when the pendulum action of mid-mounted engine weight took effect. For the sensible, though, there are no vices to worry about.

Before the Elise came along the MGF was the best-handling roadster, which means it's an extremely accomplished driver's car. Its grip in bends is simply astonishing, although there's some body roll and, unlike the Elise, the nose dives when braking.

The standard-fit electronic power-assisted steering varies its weighting brilliantly, loading up to inspire confidence in the faster bends while staying steady under straight-line running. Of the three cars, we'd say the MGF strikes the best compromise between a comfortable ride and rewarding handling. The MG also has a well controlled ride that effectively mops up the bumps.

Which leaves the MX-5. A few years back its handling was greeted with unbridled enthusiasm, but the game has moved on.

Having said that, it remains an immensely enjoyable car despite noticeable foibles. It has the softest suspension set-up of the three, and there's too much body roll. The power-assisted steering also lacks feel and is over-light. Compared with the MGF, the MX-5's handling is less precise and, driven against the Elise, it feels as vague as an American limo's.

For those who don't feel the need to wring every last ounce of performance out of a sports car though, the Mazda remains a friendly character.

Mazda MX-5 1.8i S £18,010

MX-5's handling is enjoyable, but feels woolly in such exalted company

'The MX-5 offers the best value and it's sound for everyday use'

Black plastic abounds but chrome on dials and 'eyeball' air vents hark back to classic '60s roadsters

ROADSTERS GROUP TEST

LIVING WITH THE CARS
Lotus ●●
Mazda ●●●●
MG ●●●●

You sit very, very low in the Lotus and entry is tricky because the sills you must step over are high and wide, and the seat is way down there.

Once you're in, though, the driving position is ideal. The gorgeous aluminium pedals are perfectly placed; so are the steering wheel and Vauxhall-sourced column stalks. The simple instruments relay the bare essentials of information only.

The Elise seats are thin, racing-like affairs, and an air-filled lumbar cushion is the only concession to comfort.

There's a rudimentary heater for the feet only, and vents to inhibit window misting. Tiny leather pockets and a couple of coin trays are the only designed-in cabin storage.

Behind the engine, there's a small boot that can take a large grip bag, but little else – clearly, the Elise is a weekend car.

The hood takes three minutes to erect and includes two spars that clip between the windscreen top and the roll-over hoop. Once in place, the cloth roof is rigid, though. There's a fixed glass rear window behind the driver that reduces wind buffeting surprisingly well.

Options for the Elise include metallic paint (£690), a radio fitting kit (£150), leather seats (£585), driving lights (£255) and an alarm and immobiliser (£295).

There's no spare wheel (just a tyre re-inflation kit), and a hardtop will be available soon.

The other two cars have convertible roofs that can be erected in a trice, although their plastic rear screens do mist up in bad weather. The MG has the best driving position of the three, and the most luxurious cabin. It has cream instrument dials, good switchgear, a driver airbag, electric windows and anti-lock brakes – luxuries undreamt of in the Elise, and ones that would be inappropriate in a car whose essence is its uncompromised simplicity. The MG's boot is narrow but deep.

Like the MGF, the MX-5 is a thoroughly sound car for everyday use. It has a useful boot, electric windows and mirrors, anti-lock brakes, and a fuel flap and boot-release mechanism. The cabin is dark and plasticky, but everything works well and the retro-look dials have attractive chrome bezels.

Despite its abundant zest, the Elise was easily the most frugal, returning 33.7mpg overall during our test. The MGF managed a creditable 30.2mpg and the MX-5 was thirstiest at 26.9mpg.

The Mazda offers the best value, though, having group 13 insurance and a three-year/60,000-mile warranty. The MGF is rated in group 14 insurance and has a one year, unlimited mileage warranty. There's no word on insurance for the Lotus but it has a one-year, unlimited-mileage warranty. Our guess would be group 15 or higher.

Rover MGF 1.8i VVC   £18,795

MGF's trade-off between comfort and sharp handling is well judged

'The MGF strikes the best balance between ride and handling'

MGF's cabin is the the most comfortable here, providing the best driving position

Lotus Elise v MX-5 v MGF

MGF hood is raised or lowed in seconds. MX-5 is a doddle, too, but you'll need 3min to put a top on the Elise

How good are they as everyday cars?

It's called practicality and it encompasses everything from finding somewhere to stash your sunglasses to taking the week's shopping home without accidentally making a cake-mix on the floor of the passenger footwell.

These are matters which will be furthest from your mind as you buy that roadster you always said would be yours one day. But perhaps now might just be the time to stop and ask: can I really live with one of these cars?

They are, of course, anti-social, since you are restricted not only in the number of passengers you can carry, but also the type. An elderly relative will not thank you for a lift, given the contortions necessary to get in and out. Worst by far is the Lotus because of its ultra-low driving position and the extra-wide sills you have to climb over. For women in skirts especially, the climb aboard may be undignified.

You start to take a more critical look at your luggage when you own one of these cars, too. What fits in a soft bag without creasing? What's the minimum amount of clothing necessary for

'These cars are anti-social, given the contortions needed to get in and out. Worst by far is the Lotus'

my stay away? The MG has easily the best boot of the three, but it's poor by the standards of most superminis. And because it's next to the mid-mounted engine, there's heat from the mechanicals to contend with – though not as much as in the Elise, which has a similar layout.

But most times you could just about manage with an MG or a Mazda as your only car: never with the Lotus. It's just too rudimentary; too much for driving and enjoying and not enough for living with. No proper heating and ventilation in the manner most people will have become used to; virtually nowhere to store things; no hint of the slightest concession to irrelevance in its specification.

Then, there are the hoods. While you can open or close both the Mazda and the MG in seconds with a couple of latches and a hefty shove or tug, the Lotus's takes three minutes, involving struts and stays, press studs and an Allen key. It's irritatingly fiddly rather than confusingly difficult, proof that the Elise is a weekend car rather than an only-car option. Does that deter us? Not a bit.

Elise Room for a soft bag, nothing more

MGF Deepest and biggest of trio here

MX-5 Shallow, obstructed by spare

ROADSTERS GROUP TEST

Make	LOTUS	MAZDA	ROVER
Model	Elise	MX-5 1.8i S	MGF 1.8i VVC
VERDICT	●●●●●	●●●	●●●●
Driver appeal	●●●●●	●●●	●●●●
Value for money	●●●●	●●●●	●●●
Comfort and refinement	●●●	●●●●	●●●●
Space and practicality	●●	●●●	●●●
Quality and finish	●●●	●●●●	●●●●
Equipment	●●	●●●	●●●
Safety features	●●●	●●●●	●●●●
Security equipment	●●	●●●	●●●●
ESSENTIALS AT A GLANCE			
Price	£20,415	£18,010	£18,795
Cost per mile	n/a	41.7p	53.06p
Service 3yr/36,000 miles	n/a	£1068	£933
Insurance group	tba	13	14
Test mpg	33.7	26.9	30.2
Power bhp/rpm	118/5500	128/6500	143/7000
Torque lb ft/rpm	122/3000	112/5000	128/4500
Max speed	126mph	116mph	126mph
0-60mph	6.1sec	9.9sec	7.5sec
IN GREATER DETAIL			
Running costs and warranty details			
Service interval	9000 miles	6000 miles	12,000 miles
Fuel cost (10,000 miles)	n/a	£796	£647
Anti-rust/paint warranty	8yr/–	6yr/–	6yr/–
Govt mpg (urban/56/75)	28.9/49.9/39.4*	28.2/40.9/31.0	30.4/55.6/44.6
Touring mpg/tank capacity	na/8.8gal	32.0/10.6gal	40.2/11.0gal
Warranty (months/miles)	12/unlimited	36/60,000	12/unlimited
Equipment			
Central locking	no	no	remote
Power steering	no	yes	yes
Metallic paint	£690	£210	£230
Electric windows/mirrors	no/no	yes/yes	yes/no
Radio/cassette	no	yes	yes
Steering adjustment	no	no	no
Seat height adjustment	no	no	no
Alloy wheels	yes	yes	yes
Safety and security			
Airbags driver/passenger	no/no	yes/no	yes/£345
Seatbelt pre-tensioners	no	no	yes
Side-impact door beams	no	yes	yes
Anti-lock brakes	no	yes	yes
Alarm/immobiliser	£295	no/yes	yes/yes
Deadlocks/unique-fit radio	no/no	no/no	yes/no
Space			
Length/width/height (in)	147/67/47	157/72/48	154/70/50
Track front/rear (in)	57/57	55/56	55/55
Wheelbase (in)	91	89	93
Boot capacity	n/a	5.0cu ft	7.4cu ft
Performance			
30-70mph through gears (sec)	6.1	9.8	7.7
30-50mph in 4th (sec)	5.3	7.1	6.3
50-70mph in 5th (sec)	8.1	11.2	9.7
Engine capacity/type	1796cc/16v	1839cc/16v	1796cc/16v
Kerb weight (kg)	690	990	1070
Buying essentials			
Number of colours	8	6	6
Models in range	1	3	2
Minimum PCP deposit	n/a	10%	10%
Replacement due	new car	1998	not imminent
Country of origin	UK	Japan	UK

* New EC fuel figures (see p149 for details)

All *WHAT CAR?* road tests are conducted using Mobil Unleaded Fuel or Diesel Plus containing additives to help keep engines cleaner

KEY

Price: full retail price, including cost of delivery but not road tax

30-70mph: through the gears. The best measure of performance

Cost per mile: calculated over three years, and includes depreciation, maintenance, borrowing cost, road tax and fuel, but not insurance

Service cost: includes routine servicing and tyres, brake pads and exhaust parts

Touring mpg: calculated from Govt figures by adding half of the total for urban driving to one quarter of the 56mph and 75mph fuel consumption

Fuel cost: Govt touring mpg figure is multiplied by the average price of fuel per gallon

Minimum PCP deposit: the minimum you must pay when using a personal contract plan to finance your car

For further details on all the cars in this test see the BUYER'S GUIDE starting on page 146

OUR CHOICE
Lotus Elise

The Elise is a landmark car, no question. It moves the affordable roadster into a new era, offering performance and handling of quite exceptional eagerness, and not just when measured against the rivals it faces here. It brings a taste of supercar ability within reach of a broad band of enthusiasts.

So where does that leave the Mazda and MG? The MX-5 is still a great little car and excellent value for money. No doubt there will always be buyers for it, and deservedly so.

Trouble is, the MGF outdoes it on the road and feels a more mature car for little more money. It outpoints the MX-5 comprehensively. And it's an an everyday car, something you certainly couldn't say of the Elise.

The Lotus is for uncompromising enthusiasts. It's the most rewarding of its ilk and its abilities far outweigh its disadvantages. Just try one. ■

Twin lights, twin pipes... details make the Elise a car to love

A dream to drive, but too impractical for use as an everyday car

[LONG-TERM TEST] Ten months later, **N808 SVC** is excelled in town, impressed at speed, suffered sporadic

A great sports

etched deep in Richard Bremner's heart. Our **MGF 1.8I**
electrical distress, but remains a wonderful (British) car

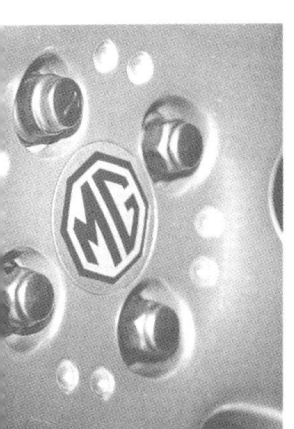

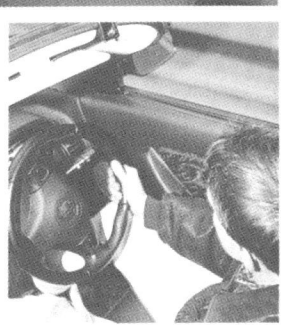

IF IT'S RAINING, I CAN NEVER RESIST it. Just half a mile from my house is a quiet left-hand, right-angle bend into a one-way street, and if I prod the throttle hard enough, the MG's tail can be provoked into a satisfying little wag. You've got to be quick on the wheel to correct it smoothly, for the F is not given to letting go, but with practice – and I've had plenty – it's very satisfying.

The MGF is good at serving up little pleasures like this when you're in the right place at the right moment but, more important, it's a gratifying drive pretty much all of the time. To start with some basics, it's dead comfortable. While some find the steering wheel too low, for me the driving position is absolutely right, the controls positioned just so and the seat (unexpectedly for it doesn't look anything special) exceptionally comfortable. I've never got out of a car, any car, after a long drive 'with nary an ache or a pain' as many a tester claims, but the damage after, say, an arduous six-hour haul, is more limited than in many cars in which comfort is allegedly a priority.

Then there's the low effort involved in driving it, which is where I'm sure Rover has learnt so much from Honda. Our 1.8i MGF came with optional electrically assisted steering that makes light work of manoeuvring (just as well, because the F's high rear deck makes it a sod to park) and complements the low-effort pedals and gearshift to great effect. Couple this with its compactness, and you have the potential for a wieldy car. Which it is. Once loosened up, the wonderfully sweet engine pulls hard enough to counter the effects of the tall gearing, and the F's swooping agility through turns is superb, even if the steering isn't as quick as might be expected of a mid-engined car. That it also hap-

car illustrated

[LONG-TERM TEST]

pens to be just about the most vice-free mid-engined car I've driven at any price (and I'm fortunate enough to have sampled Lotuses, Lamborghinis and Ferraris) provides you with a package that's more than a little alluring, and never mind the over-dainty styling, the slightly cheap interior and the sometimes noisy hood.

That a car so dynamically accomplished should have come out of Rover is heart-warming, for the company's reputation for chassis honing is not top-notch. And it's even more unusual to be saying these things of an MG, cars that were once the target of much ridicule, particularly in the pages of this magazine.

So the next question was abundantly obvious. Would this car be reliable too, a quality often depressingly absent from the products of Britain's biggest car maker?

ON DELIVERY

N808 SVC WAS NOT NEW, arriving with 4700 miles wound on during a stint on the press fleet. It was a 'quality proving car', said Rover, and in fact turned out to be the 98th off the line, which made it a little bit special in my, classic car collector's, eyes.

It wasn't quite perfect. The clock was twisted a few degrees off-centre in its orifice, the nearside corner of the boot-lid fouled the rear lamp cluster and, occasionally, there were rattles from the hood (above). Otherwise, it felt great.

THE HONEYMOON PERIOD

ONCE I'D GOT OVER THE shock of the MG's metallic purple paint (known as Amaranth, officially), a colour I eventually came to like, the enjoyment began.

That the F spent much of its life town-bound may seem a bit of a waste, but it turned out to be a great city car, because it is small, nippy and light work. The ease of lowering the hood – you can do it without getting out of your chair simply by releasing the clips and throwing the roof rearwards – meant it was often folded for the short trip to work. The only drawback is that you ingest the fumes of fellow road-users, and rapidly learn that the main sources of pollution are diesel-fuelled buses, trucks, vans and the odd badly maintained car. The only source of irritation from the F itself was the difficulty of negotiating the ebb and flow of traffic without the powertrain shunting when the throttle was released. This is a problem that Rover, BL and BMC have had with transverse-engined cars since 1959, and it seems astonishing that it cannot banish it – almost every other car maker has.

In town the MG collected admiring glances from passers-by and, out of town, growing admiration from its user. Driving with more zeal revealed the twin-cam four's throatiness, the accuracy of the gearchange, the strength of the brakes (whose pedal weighting and firmness is just right) and the car's fantastic grip and balance. On dry roads, it was virtually impossible to make it let go and on wet roads slides were difficult to achieve unless the surface was greasy. It always felt wonderfully secure. And, so far, it seemed wonderfully dependable.

TROUBLE WITH GLASS

I DIDN'T NOTICE IT UNtil it had grown a couple of inches, but there was a crack in the windscreen that travelled up from the demister before turning left to head for the centre of the car. It looked like a stress fracture, there being no stone chip in sight, which surprised me, because the MGF does not suffer much scuttle shake. The next day, I noticed another crack, growing from exactly the same spot on the opposite side of the screen – spooky. Over the next few days the cracks travelled towards each other, and I booked the car into Muswell Hill dealer Palmsville, hoping the screen wouldn't drop into my lap like a shattered dinner plate.

Palmsville were to replace the screen, and I also ordered a new nearside door trim because somehow, the tongue of the seatbelt had punched a hole through it when the door was shut. I told Palmsville I would fit that myself, figuring on saving some money.

They had the car overnight, for the screen-bonding glue to cure, and explained that the cracks were caused by defective glass rather than bodywork problems. They fitted the door trim free of charge, a great gesture, and washed the car too. Problems over, I thought.

Until the cracking sound. Every time I hit the button to lower the driver's side electric window there was a sharp crack before the glass de-

> 'That a car so dynamically accomplished should have come out of Rover is heart-warming'

scended sillwards. The glass wasn't breaking, but sounded like it wanted to. A couple of weeks later, the pane would not ascend without a helping hand, and I knew I had an unscheduled visit to an MG dealer in prospect. There are not many of these in the capital: the nearest was miles away and I didn't fancy the crawl through traffic to get there. So I lived with the problem until the 12,000-mile service. Keens of Battersea got the job, and besides fixing the electric window I also asked them to sort an overkeen handbrake warning light. I forgot to mention that sometimes the roof would jam, half-folded, when you tried to lower it. Opening and shutting it a few times usually produced a submission, and I'd forget about it again. It was only in the last week of ownership that I discovered that one of the small struts at the front of the hood-frame was going overcentre. Keens had the car ready on the same day, but it had not been cleaned, nor had its service book been stamped. On the other hand, the bill was £92.70 – less than for any other Rover, including the 100, listed on their 12,000 service menu.

PILING ON MILES

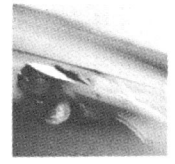

THE MILES PILED ON AT quite a rate. A long weekend in southern France helped, though it didn't do a lot for the eardrums – the MG is noisy at high speed, and the radio, fulsome at a moderate pace, is not up to the battle at 85mph, hood up or down. Hot weather showed up the inadequacy of the ventilation system – and the noisiness of the fan – but then, folding the lid rearwards instantly fixed that problem. Despite suffering the pounding of London's pot-hole ridden streets, the MG felt as taut as ever bodily, even if the hood kept on rattling and the lid under which the oil filler lurks kept on zizzing.

By now I'd come to admire the MG's practicality as much as its dynamics. The boot was big enough for a weekend away for two, I liked the painlessness of spending hours in it and the sheer ease of driving it. I also liked the illuminated ignition key slot, the delayed dimming of the cabin lights, the accessibility of the dipstick (and the built-in dipstick wiper, though that seems to get quite oily), the cabin's cubby-holes and the fact that, by directing the centre vents at the plastic rear window and turning up the fan, you could swiftly demist it.

I didn't like the fact that the remote control door mirror handles broke off in my hand, that the tonneau cover is absolutely useless, being

[LONG-TERM TEST]

LOG BOOK
DURATION OF TEST: 10 MONTHS
MILES COVERED: 15,563
PRICE NOW: £17,295 THEN: £15,995
ENGINE: 1796CC 16-VALVE FOUR
POWER: 120BHP @ 5500RPM
TORQUE: 122LB FT @ 3000RPM
OVERALL FUEL
CONSUMPTION: 33.3MPG
MAXIMUM SPEED: 120MPH
0-60MPH 8.5SEC
SERVICE COSTS: £92.70
REPLACEMENT COSTS: £128.60
TAX: £117 INSURANCE: £465.35
FUEL: £1005.75
OIL BETWEEN SERVICE: £3.50
DEPRECIATION: £995
COST PER MILE
EXC DEPRECIATION: 11.6P
INC DEPRECIATION: 18.0P

Vandalised bonnet and cracked bumper (below), which had to be replaced

Controls pretty convenient, but watch the fuel gauge – the needle goes nowhere for 250 miles, then plunges earthwards. Radio fair

impossible to fit and poorly made, and that the tonneau-retaining stud inside the nearside door jamb was rubbing paint off the door itself. Nor that the rubber buttons in the remote locking key fob were sinking into the unit itself.

But the mechanicals were in great shape, the 1.8 K pulling with noticeably more urgency at 15,000 miles. I found myself driving the MG almost to the exclusion of other test cars.

The MG's fortunes took a turn for the worse in August, when an MR2 tapped the back of it, and vandals scored the bodywork. The rear bumper was irreparably cracked, the number-plate broken. Result: a bill for £384.39. Not much protection for the wallet from low-speed shunts there, then. And, the remote key fob had stopped triggering the locks. Prising it apart is a nail-breaker, but once done, I could push the rubbers back into place. The fob is made by Lucas, which can be cursed just like it has been by MG owners past. More electrical trouble struck a while later, when the alarm took to hooting uncontrollably. The radio died too, and the boot and cabin lights no longer illuminated either. All are on the same circuit. Attempts to resynchronise the alarm failed to halt its bleatings, so I rang Rover's customer services and bleated in revenge. They were sympathetic, but could only suggest visiting a dealer, which was not convenient. So I disconnected the horns.

Cold weather meant the hood was now rarely lowered, but driving this soft-top in winter proved no bind. The heater is effective enough to make the cabin snug (though the fact that you can't have warm feet and cool air through the vents is a pain) and the hood is watertight. But not if you give it the bucket and sponge treatment, curiously. Enthusiastic sponging also destroyed two side repeater flashers whose lugs break if they're struck firmly by a sponging hand. They cost three quid a go.

Palmsville had a go at the bodywork, achieving a good colour match, but filled some bootlid dents badly and left dirt in the paint. They replaced the key fob casing, sorted the alarm and a faulty airbag warning light and lubricated the hood frame to allow it to fold, for £128.66, the only extraordinary bill.

CONCLUSION

AFTER 10 MONTHS AND 15,000 miles, I have real admiration and affection for the MGF. Steering is not the most sensitive, but the chassis's wieldiness, balance and ride quality, the engine's enthusiasm and its mechanical and structural integrity make it a pleasure to live with, despite minor glitches. That the F was developed on a shoestring employing little more than blind enthusiasm and Metro bits makes this a very fine achievement. The hood could be better finished, the interior looks a little cheap and it has not been fault-free – but the MGF's fundamentals are very well sorted.

Apart from being notable for its abilities, this MG will also go down in history as the last all-British car to be built by the company that was once BMC. How fitting, then, that it should be a descendent of the Mini, that company's greatest car, and that it should work so well.

'Rover gave him a veneer kit as compensation'

CAR READERS TELL US WHAT IT'S LIKE TO LIVE WITH THE MGF

No doubt about it – all the readers we contacted liked their MGFs, including the ones who had problems with their cars.

BARRY GARTNER, WHO HAS A 16-month-old 1.8i, reckons, 'On a summer's day with the hood down and the radio on it's a really nice car. You can't really fling it around in south-east England, but it's wonderful hitting a roundabout and getting it right. And the ride is stupendous. 'The MG is Gartner's first sports car, chosen over a Fiat Coupé, whose looks he considers 'a little dated – the MG is more classical'.

Any problems? 'That's the downside. There's been no trouble with the engine and suspension, but there has been with the body and hardtop. It's spent a bit of time at the dealer.' A rattling, cracking hardtop was the trouble. A second lid, from a different supplier, has been better – after the door windows were adjusted and modified seals were fitted.

Gartner has also had the door mirror remote handles break, the hood zip fail – there's now a stouter replacement – rattling door lock buttons, a failed heater fan and a bouncing speedo needle. He also had to wait ages for the hardtop to be delivered, but Rover gave him a wood veneer interior trim kit in compensation (!). He also feels that the one-year warranty is inadequate. Despite all this, he'd buy another.

ANDY RAKE, BY CONTRAST, HAS had no problems whatever with his VVC, which he considers 'absolutely superb'. Likes include 'the flexibility of the engine, the superb handling and the very impressive ride. I had a 16V Calibra before, and you could tell that the suspension was working uncomfortably hard at 65mph. In the MGF, it's more like 85-90mph.' Rake has a hardtop, and reckons it 'absolutely transforms the car – it's so civilised'. Dealer Wrights of Lincoln are 'very good'.

ROBERT COLLINS' MGF VVC shares a garage with a BMW M3, and has 'pleasantly surprised him.' He compliments the 'superb ride which, compared with the M3, is very compliant without losing any roadholding', the rigidity of the body and the more than adequate power. He is less keen on the rattling hardtop, but says the dealer is slowly sorting that out.

The same dealer also took three attempts to respray the chipped bonnet before getting a good colour match, and had to replace the hood zip. He's also had trouble with the same electrical circuit as CAR – the fuse covering the interior lights and the radio keeps blowing. Otherwise, he's 'pleased with the purchase', considering it 'good value', but he wouldn't buy another, fancying a BMW Z3 2.8 instead.

JOHN WHARTON REPLACED AN MX-5 with his MGF 1.8i and bought a car from Rover 'with some trepidation'. But the MG has been 'excellent', the only trouble being wear of the inside shoulders of the front tyres, which were replaced by the dealer, who also reset the front end alignment. Wharton considers the MG's interior '100 percent better' than the Mazda's, and the whole car an advance except for the gearchange – he misses the riflebolt precision of the MX-5's shift. Dealer Horners of Eccles are excellent – they collect and deliver the car, which is a 100-mile round trip. Wharton would definitely have another MG.

SO WOULD **BARRY CARPENTER**, although he plans to keep his VVC for a long time.'I'm delighted with it.' This, in spite of getting to see his local AA man twice – once, to re-attach the gearlever on the day the car was delivered, and once at 1100 miles to diagnose a failed alternator. Otherwise, there's been no other trouble, and he's pleased with the service from Exeter Motors.

MG FORCE

ARE YOU TROUBLED BY THE THOUGHT THAT ANTHEA TURNER DRIVES AN MGF? ARE YOU WORRIED THAT EVERY MGB YOU SEE IS DRIVEN BY A MIDDLE-AGED BLOKE IN A CLOTH CAP? WORRY NO MORE - THESE TUNED VERSIONS ARE SERIOUSLY ENJOYABLE

Words: Daniel Strong. Pictures: Michael Bailie

Ask any MG owner why they've chosen one and they'll probably spit squashed flies all over you as they rant on about them being the quintessential British sports car. Ask any hardcore driver what they think of MGs and you'll get a different, and rather less charitable story.

Even when new, the MGB suffered from lazy handling, a wheezy engine and below par performance. Sure, it had loads of character and got the cloth cap brigade in a lather, but any driver worth his stringback gloves kicked the B into touch and bought a Lotus Elan.

Haunted by the inspirational Lotus Elise, the MGF is also suffering something of an image problem, gaining the reputation of being a bit of a girl's car. Is there any hope for either of them?

To find out, I've come to respected MG specialists, Brown & Gammons. This Baldock-based company has an unrivalled reputation for preparing MGBs for historic rallying, and for fettling road cars into serious sports cars. Parked outside is a striped, sweet and thoroughly sorted rallying 'B', ready for me to drive. And to see how this 1960s swinger stands up to modern machinery, I've brought along a Moto-build modified MGF. Well known for breathing on the Rover Group's more mundane products, this is Moto-build's first attempt at toughening up the 1990s MG.

I've always been a closet fan of the slovenly but charismatic 'B'. More so than the 'F' if I'm honest. In the past, no-one's been able to convince me the 'F' is a real good 'un, or that it has the character of its predecessors. Clearly both modified cars have a lot to prove.

Both Moto-build and Brown & Gammons are out to show that their cars are red-blooded sports cars, and although the two companies deal with cars of very different ages, their methods are similarly modern and effective. Each has worked on the suspension and brakes, as well as the engines, although clearly the

MG FORCE

unsophisticated old 'B' has more room for improvement. And because age comes before beauty (so wrinkly Editor Fraser says, anyway), let's look first at the MGB.

Despite a modern makeover, the Brown & Gammons 'B' still feels and smells old. But it isn't all like it used to be. The interior's been completely revised. FIA-approved race seats and four-point harnesses hold you in place, while a complex roll-cage has been mounted behind you and fire extinguishers bolted in up-front. It's not essential gear for road use, but it makes you feel good when you strap yourself in. Outside, the conventional hard-top has been replaced by a taller, works-style version, bolted down to reduce body flex. Clamber in and you're immediately struck by just how low you're sitting. Your legs are stretched almost flat along the floorpan, and with the Motolita wheel close to your chest, you feel like Stirling Moss.

It's certainly a world away from the plush cream Connolly leather interior of the MGF, with its airbag, stereo and custom door-trims. While the 'B' makes you feel like a racing driver, the 'F' makes you feel like a wealthy property developer. It cossets you in a way the 'B' could never do, rally car or not. The whole feel is modern, but although

Moto-build has covered the MGF's seats and door panels with Mr Connolly's finest creamy leather. Very nice

Moto-build's decision to let Connolly trim the seats and door panels was a good one, the 'F' is still a bit plasticky and a bit too 'normal'.

Getting into the 'B' is an unnerving experience. Old cars tend to be hard to drive – and I'm not expecting Brown & Gammons' £17,500 MGB to be any exception. The gear lever needs teasing through the four speeds of the straight-cut 'box (overdrive on third and fourth makes things a little easier), and the unservoed brakes need serious pedal pressure before the pads bite into the grooved discs. Still, at least they've got some bite, unlike the gummy stoppers of the standard 'B'. The Moto-build MGF's brakes, on the other hand, are spot-on. Thirty years of progress shows up pretty clearly in the

Racing seats and harnesses, Motolita wheel, plain black trim... Brown & Gammons' MGB means business

braking department, and the Kevlar composite pads haul the MGF up strongly. You might wish for the ABS option on wet or frosty days, but there's no drama if you're sensible.

Tyres have come a long way, too. The MGB's lovely wire wheels are shod with replica '60s forest rally tyres, which don't exactly fill you with confidence when you're trying to drive down a wet, winding country road on a cold December morning. Give them and yourself a chance though, and they hold no scary surprises, but they don't have the immediate, confidence-building grip of the MGF's gumball Fuldas.

The 'F' isn't all good news, though. It has always had slightly woolly steering about the straight-ahead, and the

MGB (right) has tweaked 1.8-litre 'four'. Hairy cams, bucket-size carbs, competition exhaust and lightened and balanced internals give it an extra **45** horses and the fruitiest exhaust note. (We'd show you the **MGF**'s engine, but you have to remove it from the car)

'The F's a bit snappy when the back starts to slide. Keep your wits about you and it's challenging, sweaty-palmed fun'

MG FORCE

wider TSW Blade alloys and 205x45 tyres do it no favours. The lowered and stiffened suspension, which is 1.2in down on standard and a fair bit harder, firms things up nicely, but the steering is still disappointingly uncommunicative. Things improve after the first half turn of lock, but it's still a long way from the surprisingly direct helm of the MGB.

The B's ride isn't as hard as you'd expect but it's still much stiffer than the standard car. Body-roll has been reduced with thicker anti-roll bars, uprated dampers and special Brown & Gammons springs, and as a result every crest launches the car into the air. It's never going to be faster than the MGF point-to-point, but once you've settled in you can start to have some fun. You have to drive virtually flat-out to get the best from it, but as you're not travelling at lunatic speeds, this isn't as daunting as it sounds. Turn in normally and it'll understeer gently. Throw it at the inside verge 10ft away from the corner, lift off and the whole car will drift gently. Once you can see out of the corner you can nail it, hang the tail out, sing a rousing verse of Jerusalem, and almost kid yourself you're the new Stirling Moss.

That's something you can do with the Moto-build MGF, too, although don't expect such a forgiving chassis. The extra mid-range grunt created by the mods to the engine means power oversteer is a very real possibility. But because of the mid-engined layout it's all a bit snappier when the back starts to slide. The bigger, lower profile tyres have increased the F's roadholding, which is no mean feat, but this means you're going quicker when it does let go. Keep your wits about you and it's challenging, sweaty-palmed fun. Hell, after a few miles you might even sprout a few hairs on your chest.

The engine bays of both our MGs contain surprises, too. The B's engine has been fettled the old-fashioned way, bored-out to 1840cc from a shade under 1800cc, with all the internals lightened and balanced. Bigger carbs fill the cylinders with more fuel, and the gasses rumble out of a stainless steel tail-pipe at 100Db – *Bwaaarp!* Moto-build, as is the modern way, has called in the good Doctors of Technology for a dose of added power. With a remapped engine management system courtesy of Superchips, hotter cams and Pipercross air filter, it has given the basic 1.8 K-series engine as much power as the all-singing VVC unit.

Blip the throttle and you get the added bonus of an induction whoosh and, crucially, a similarly throaty *Bwaarp!* The noise is so fruity you could eat it. Rover neglected to give the 'F' a proper rasp, and although its absence was noted, I don't think anyone could have anticipated just how much a true MG depends on it. The twin three-inch pipes of the Moto-build system help the 'F' wear its famous octagon with pride.

Chasing along the road in close formation, both cars bark at each other noisily. You can hear the 'B' from the cockpit of the 'F', and vice versa. It's almost a conversation in MG, as one car echoes the other. And it's at this point that the 140bhp Moto-build 'F' really comes into its own. It whips past the 125bhp 'B' and off into the distance before you've flicked the old-timer's 'box out of overdrive. The mid-range urge of this 'F' is actually greater than the VVC,

Above: modified 'B' is a great car in which to polish up your Stirling Moss impersonation. This is the four-wheel drift

Separated by 30 years, but united by a common thread: they're bloody great to drive. (Is that common enough for you?)

64

Specifications	MGB	MGF
Engine	Four-cylinder, in-line	Four-cylinder, in-line
Location	Front, longitudinal	Mid, transverse
Displacement	1840cc	1796cc
Maximum power	125bhp @ 5000rpm	140bhp @ 6100rpm
Maximum torque	115lb ft @ 3000rpm (est)	142lb ft @ 2800rpm
Transmission	4-speed (overdrive 3rd/4th)	5-speed
Front suspension	Independent, wishbones, coil springs, anti-roll bar	Wishbones, Hydragas springs, anti-roll bar
Rear suspension	Beam axle, coil springs	Wishbones, Hydragas springs, anti-roll bar
Brakes	(front) grooved/vented discs (rear) drums	(front) vented discs (rear) solid discs
Wheels	72-spoke, 'works' replica wires	TSW Blades, 7.5 x 16
Tyres	Dunlop SP Sport, mud and snow, 195/65 x 14	Fulda Y2000 205/45 x 16
Price	**£17,500** Brown & Gammons, tel: 01462 490049	**£18,500** modified (£16,395 list price) Moto-build Racing, tel: 0181 893 4553

and on the road you can feel that. The engine pulls strongly from 2500rpm and gives you a kick in the back as it launches into the power-band at 5000rpm before the limiter calls a halt at 7000rpm.

But as the 'F' disappears over the horizon you know you're getting as much enjoyment out of the 'B'. You can't expect an MGB, however well sorted, to keep up with a modern sports car, but somehow it doesn't matter. The great thing about the Brown & Gammons car is that it's still got the traditional appeal of a true MG, but with enough ability to make it rewarding rather than frustrating to drive hard.

Although the fully converted 'B' is not as powerful as the 'F', it's still a bit too extreme for everyday use. It's a fun weekend car, not something for serious commuting. Harnesses and full race seats are a bit of a bind, and you need to heel and toe, and double declutch constantly if you're to change cogs without a hair-raising *graunch*. Brown & Gammons promise they could build a car similar to the rallying 'B', but without the racing seats and gearbox. That would be a lovely, lovely road car, and a bit cheaper than the modified 'F'.

Moto-build's MGF, finely honed and far more aggressive than the standard car, is still far more user-friendly than the 'B'. It's the kind of car you can drive long distances, wring out all the performance, and still arrive at the other end unruffled but grinning from ear to ear.

So, more than 30 years after the launch of the 'B', it would seem that far from being also-rans in the sports car stakes, MGs can be turned into really great driver's cars. The only problem I can see is deciding just which one to choose!

BMW Z3 v Mazda MX-5 v MGF

Fun boy

Late, perhaps, but BMW has timed the arrival of its Z3 perfectly for the British summer. Can the MGF and MX-5 cope?

Eighteen months is a long time to wait for any new car, but it must seem interminable when a dealer is already holding your deposit. Yet that's what 3306 eager Brits have been prepared to tolerate to get their hands on a BMW Z3.

Now, at last, they're being delivered to UK customers, and to prove that patience has its reward, the price originally quoted has been frozen – just over £20,400 on the road – though anyone joining the year-long waiting list now is likely to have to pay another five per cent or so.

That sort of price frame puts the Z3 into almost the same ball park as the MGF and MX-5, the car that created a market for an affordable roadster at a time when no other manufacturer had even sniffed the possibility.

Zappier Z3s will take on the Porsche Boxster and Mercedes SLK before the end of the year, but the only version currently available is this 1.9-litre.

On intent, specification and performance it couldn't be closer to an MGF 1.8i VVC or Mazda MX-5 1.8i S. Those 3306 Britons currently drooling over a Z3 delivery date may not wish to read on, but others still thinking about taking the plunge should, perhaps, risk scanning through to the end of this test.

CONTENDERS AT A GLANCE

BMW Z3 1.9, £20,420
BMW's US-built roadster is here. Bad news is that there's a long queue

Mazda MX-5 1.8i S, £18,345
Car that sparked the roadster revival. A little crude now, but still huge fun

MGF 1.8i VVC, £19,795
First and cheapest of the mid-engine two-seat drop-tops. Smooth and supple

ROADSTER TEST

three

ROADSTER TEST

BMW Z3 1.9 £20,420

Perfectly weighted steering; facia too similar to Compact's

Delightfully traditional roadster profile; bigger engines to come

'The beautifully reassuring Z3 refuses to be budged from its course'

Higher equipment levels than MG or MX-5; widest range of seat adjustment. Worst fuel economy; very secure handling but lacks the sharpness of Mazda

SHOWROOM APPEAL	
BMW	●●●●
Mazda	●●●
MG	●●●●

In function if not form, these cars have strong similarities. All are two-seaters with manually folding roofs. They're close in size and each has a four-cylinder 16v engine of less than 2.0 litres driving its rear wheels.

But it's the varying way these elements have been packaged that gives each car its distinctive appeal. The BMW and Mazda engines are mounted under the bonnet in traditional roadster fashion, but the MG's is housed transversely behind the passenger cabin, supercar-style.

That gives the MGF a high and long boat-deck tail and short, low nose. The BMW is the reverse – long-fronted and aggressive in anticipation of its large engine bay swallowing the 2.8- and 3.2-litre six-pot engines still to come. It's not the proportions that spoil the look of the Z3 but the colander-style wheels. You'd never believe they're proper alloys.

All three cars hark back to the past, most notably the Mazda which has the petite and well-balanced looks of the original Lotus Elan. The BMW has a row of fake heat exhalers in its front flanks behind beefy wheel-arches; and the MG is lifted by meshed side air scoops and a beautiful nuts-and-bolts-looking filler cap.

DRIVER APPEAL	
BMW	●●●●
Mazda	●●●●
MG	●●●●

Rarely does a BMW fail to deliver a glow of satisfaction, so prospects for the Z3, even in this tamest form, looked good. Yet it's the MX-5 that serves up more fun in short blasts than any mainstream rival bar the Lotus Elise.

The Mazda may be the slowest and inflict the stiffest ride, but find a well surfaced, winding road and you'll be roundly entertained. It's an old-school roadster, comparatively low on grip, but delightfully informative and responsive to the very direct steering. You'll also revel in its hair-trigger throttle responses, and the short, swift action of the gearshift. It's outstanding, even by the high standards of the BMW and MG shifts.

Outputs of the Mazda's 1.8-litre engine – 128bhp and 112lb ft – are modest, but low gearing and the MX-5's lightness keep things sharp. Being a 16-valver devoid of the cylinder-head trickery of the BMW and MG, though, you need to work its middle gears hard to defeat low-rev lethargy.

A bigger problem is the relative whippiness of the MX-5's chassis. It contorts untidily on bumpy country roads that twist and undulate at the same time, and there's more jitter and judder to its ride almost everywhere.

68

ROADSTER TEST

Mazda MX-5 1.8i S £18,365

Responsive steering a joy; dash looking dated; best gearshift

Another trad body style; technically the simplest car here

'The Mazda is more fun than any roadster short of a Lotus Elise'

Seat cushion too short. Delightful handler because grip levels are lowest; so enthusiasts can explore limits in greater safety. Boot the least satisfactory

◁ You don't get that in the BMW; nor should you, considering it weighs almost 300kg more – a handicap reflected in the worst fuel consumption figure (31.5mpg) over our touring route. The Z3 is also less sharp than the MX-5 in every other way, and its engine note is less interesting than the slightly coarse snort of the Mazda or lovely bellow of the MG.

But the Z3 has a beautifully reassuring feel; it simply refuses to be budged from its course. The steering is perfectly weighted and its suspension is firm enough to check body movement; yet it still stifles shake or tremor most of the time, and the tyres have more grip than the engine's 133lb ft of torque and 140bhp can dislodge.

The BMW and MG motors use differing fuel induction magic to ensure that touring flexibility does not compromise outright performance, and vice versa. In the BMW, the length of the pipes feeding in the fuel/air mixture varies to suit driving styles, whereas the MGF's 1.8 VVC unit adjusts its valve-gear operation to achieve the same end.

Both engines are generally smooth (apart from some low-speed hesitancy in the Z3), easy-natured and potent. Even so, the MG's is something special. You can rev it to the hills – 143bhp of peak power arrives at 7000rpm – yet much of its 128lb ft of pull is available low in the rev range to ensure effortless pick-up. It shows in the comparative figures. The MG posts the highest top speed and quickest 0-60 time. Of more real-world relevance, it's also the quickest from 30-70mph so needs less overtaking space than the BMW or Mazda.

What the MG isn't, in any way, is sporty. It has supermini-light pedal actions; rolls the most and is the first to run wide if pitched hard into turns, though it tucks in tightly and safely when you back off. Even the ride is supple, although it's disturbed by ridges and potholes. And the steering – which varies its assistance according to speed and load – lacks sensitivity. It's too soft when you're running in a straight line, and too vague when you start to feed in some lock. The MGF is a refined two-seat tourer, not a seat-of-the-pants racer.

BEHIND THE WHEEL
BMW ●●●●
Mazda ●●
MG ●●●

By today's standards, all three are budget roadsters. That much becomes apparent when you slip behind the wheel. None of the cabins quite exude the glamour suggested by the cars' exteriors.

The Rover effort is best. At least it has tried to liven up the MGF's interior with cream instrument faces, maroon needles and octagonal MG badges wherever they will fit. Even so, the dash is more mainstream hatch than one-off roadster.

But the MG facia is a more

69

BMW Z3 v Mazda MX-5 v MGF

MGF 1.8i VVC £19,795

Most attractive facia; assisted steering lowest on sensitivity

Mid-engined layout gives MG the most up-to-minute profile

'Most supple and refined car here, MG also has the best engine'

Cabin looks like average hatch's; roof folds quickest; no seat-height adjustment. Lovely engine noise; rolls most in bends, but is fastest car and best overtaker

welcoming sight than the flat-topped affair in the Z3, which is too close to the three-door Compact's on which much of the car is based. Worse still is the Mazda's interior – a retro-look trend-setter when it was launched, it now just feels dated.

Best-equipped, surprisingly, is the BMW. It has the essentials of both rivals, including major safety and security items, plus a couple of things they leave out. Most important is seat height adjustment, one of the factors that make the Z3 the roomiest car here. It also has the most seat travel but, like the Mazda and MG, no steering column height adjustment. The MX-5 and MG seat heights are fixed, too, limiting kneeroom if you're tall.

The BMW's seat is the most comfy. The backrest may be a bit narrow for some, but for most it's better shaped than the Mazda or MG chairs. The Japanese car's cushion is too short and the Brit's too high, although the shape of the seat itself is hard to fault.

The BMW and Mazda cabins are the most blustery when the roof is down, mainly because of the way the wind creeps around the side windows (lowering them in all three cars has little effect on turbulence but does raise noise levels). And only the BMW offers an optional wind deflector.

Folding the roof in all three cars involves unlatching a couple of catches and pushing the entire structure backwards – the work of a few seconds. Fitting the tonneau covers takes a little longer. The BMW's is strongest and neatest, the Mazda's the most fiddly. The MG's is quickest to put in place but most untidy once the job is done – it perches on the back like an old sack. None of these cars has any form of rear window demisting.

PRACTICALITY	
BMW	●●●●
Mazda	●●
MG	●●●●

The MG has the biggest boot – surprising, perhaps, given its mid-engined layout – but the BMW isn't far behind; and the shape of the Z3's luggage space helps it cope with more awkwardly shaped loads. It is wide and long but shallow, whereas the MG's is tall and slim.

Both, however, are better than the Mazda's, which is neither one thing nor the other and cluttered with a space-saver spare wheel.

The BMW is best for in-car living space. It has door pockets not found in either rival, and a couple of small, lockable panniers between the seats. The MG is the only one with seat-back map pockets, though reaching them is a challenge, and there's a usefully sized cassette box next to your elbows.

The Mazda has the best glovebox, but not much else. It does have remote boot and fuel-flap releases not found in the BMW and MG, however.

70

ROADSTER TEST

Make	BMW	MAZDA	MG
Model	Z3 1.9	MX-5 1.8i S	F 1.8i VVC
VERDICT	●●●●	●●●	●●●●
Driver appeal	●●●●	●●●●	●●●●
Value for money	●●●	●●●	●●●
Comfort and refinement	●●●	●●	●●●
Space and practicality	●●●	●●	●●●
Quality and finish	●●●	●●●	●●●
Equipment	●●●●	●●	●●●
Safety features	●●●	●●●	●●●
Security equipment	●●●●	●●	●●●●
ESSENTIALS AT A GLANCE			
Price	£20,420	£18,365	£19,795
Cost per mile	n/a	42.6p	53.1p
Service 3yr/36,000 miles	tba	£1068	£933
Insurance group	14	13	14
Test route mpg	31.5	40.3	51.2
Power bhp /rpm	140/6000	128/6500	143/7000
Torque lb ft /rpm	133/4300	112/5000	128/4500
Max speed	123mph	116mph	126mph
0-60mph	8.4sec	9.9sec	7.5sec
IN GREATER DETAIL			
Running costs and warranty details			
Service interval	9000 miles (est)	6000 miles	12,000 miles
Fuel cost (10,000miles)	£936	£732	£576
Anti-rust/paint warranty (yrs)	6/–	6/–	6/–
Govt mpg (urban/56/75)	24.8/47.1*	28.2/40.9*	30.4/55.6/44.6
Touring mpg/tank capacity	35.3*/11.2gal	31.0*/10.6gal	40.2/11.0gal
Warranty (months/miles)	36/60,000	36/60,000	12/unlimited
Equipment			
Central locking	remote	no	remote
Power steering	yes	yes	yes
Auto gearbox	£1240	no	no
Electric windows/mirrors	yes/yes	yes/yes	yes/no
Metallic paint	£360	£250	£240
Steering adjustment	no	no	no
Seat height adjustment	electric	no	no
Air-conditioning	£1100	£1395	£1200
Leather trim	£850	£923	£550
Safety and security			
Airbags driver/passenger	yes/£480	yes/no	yes/£355
Seatbelt pre-tensioners	yes	no	yes
Side-impact door beams	yes	yes	yes
Anti-lock brakes	yes	yes	yes
Alarm/immobiliser	yes/yes	no/yes	yes/yes
Deadlocks/unique-fit radio	yes/no	no/no	yes/no
Space			
Length/width/height (in)	158/67/50	157/72/48	154/70/54
Headroom	excellent	excellent	good
Legroom	excellent	excellent	excellent
Boot capacity	6.3cu ft	5.0cu ft	7.4cu ft
Turning circle	32.8ft	29.9ft	34.4ft
Performance			
30-70mph through gears (sec)	8.2*	9.8*	7.7*
30-50mph in 4th (sec)	8.0*	7.1*	6.3*
50-70mph in 5th (sec)	11.1*	11.2*	9.7*
Engine capacity/type	1895cc/16v	1839cc/16v	1796cc/16v
Kerb weight	1260kg	990kg	1070kg

*Figures obtained at Millbrook Proving Ground using Econotest equipment

Buying essentials			
Number of colours	10	6	6
Models in range	1	3	2
Replacement due	new car	spring '98	not imminent
Country of origin	USA	Japan	UK

*New Euro-average consumption figures

All WHAT CAR? road tests are conducted using BP Unleaded Fuel or BP Diesel Plus with additives to help keep engines cleaner

KEY

Price: full retail price, including cost of delivery but not road tax

30-70mph: through the gears. The best measure of performance

Cost per mile: calculated over three years, and includes depreciation, maintenance, borrowing cost, road tax and fuel, but not insurance

Service cost: includes routine servicing and tyres, brake pads and exhaust parts

Test route mpg: Measured using the *What Car?* 70-mile route, including motorways, A- and B-roads

Fuel cost: Govt touring mpg figure is multiplied by the average price of fuel per gallon

Minimum PCP deposit: the minimum you must pay when using a personal contract purchase to finance your car. (PCPs are a variation on hire purchase giving low monthly repayments).

OUR CHOICE
MGF 1.8i VVC

Combine the best talents of these three cars and it's possible to distil the near-perfect lower-priced roadster. You'd add the potent, ever-willing, musical engine of the MG to the lively chassis feel and reactions of the lightweight Mazda, then stir in the looks and cabin space of the BMW (but not the wart-effect leatherette which is one of the standard trim options).

Sadly, such mix 'n' match combos are not an option, so compromises are necessary; in which case, our money goes on the MG.

It is not, we'll admit, a full-blooded sports car. It handles, rides and steers too softly. But it's a better car to live with and more forgiving of rear-drive novices than the Mazda; and it's not an adaptation of a budget hatch like the Z3. It's also fast, well-priced and desirable, all of which will help values. Unless you're a purist, you'd have to agree. ∎

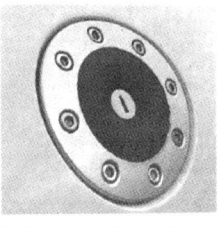

Long tail of mid-engine format   Pretty retro design touches

All three have strengths, but MGF is most complete roadster

MGF 1.8 VVC FINAL REPORT

Purple pros, purple cons

The MGF has proved to be a capable and user friendly daily motor, but after 20,000 miles Allan Muir's initial expectations have not been entirely fulfilled

AUTOCAR LONG TERM

Good looking, quick, British; it's easy to see why 11,000 have been sold

Until our long-term MGF arrived, the only time I had spent more than two consecutive days in a drop-top car was on a week-long thrash around Europe in a Caterham 7, which was jolly good fun but hard work. Roadsters struck me as being great weekend toys but rather impractical as everyday transport. However, the MGF promised to advance the roadster cause big time with a combination of driver appeal and genuine user friendliness.

After a year and 20,000 miles in our wacky Amaranth-coloured MGF 1.8 VVC, we are still full of admiration for Rover for building this car. The MGF is quick, rides well, looks stylish, doesn't leak, has a good-sized boot, and best of all it's British and wears an octagonal badge. No wonder 11,000 have been sold since its launch in autumn '95, with buyers queueing for months in the early stages to get their hands on one. It deserves to be a success, and it is.

The ongoing demand for MGFs has had the effect of propping up residual values to an impressive degree. Our VVC-engined car cost £18,795 in May last year (or £20,560 with the optional hard top, passenger's airbag and pearlescent paint taken into account); a year later it still has a trade value of about £17,500. As a result of this, and the fact that the car has required only one service in 20,000 miles, the actual running cost over a year has been very low at 24.7 pence per mile (excluding insurance), or 9.5 pence if you ignore depreciation.

Strangely, though, my opinion of the MGF has faded over the course of a year. It arrived just in time for summer, so there was an extended honeymoon period when I took every opportunity to drop the roof and use the car as God intended. But during the last six months of its time with us I became increasingly irritated with it. Partly that was down to the car and partly it was due to my realisation that I'm not actually a great fan of open-top motoring.

Part of the problem is that, at 6ft 1in, I don't fit into the MGF very well. The cabin is quite narrow, which means there's barely enough space between my right elbow and the door. With a relatively high-set seat and a fixed steering wheel, the driving position is less than ideal – my left leg touches the bottom of the wheel most of the time. And the fact that I sit high means I have to duck down to see traffic lights under the header rail when the roof is raised.

Ultimate destination in Europe was San Marino, for the GP

Although we've had several reports of reliability and quality problems from other MGF owners, our car has given us little cause for concern. From the early days the facia has generated one or two rattles (most noticeable with the roof up), and an attempt to cure a rattling hard top by fitting a new rubber seal worked only temporarily. In addition, the fuel filler cap lock has jammed on more than one occasion, culminating in the lock actually pulling out of the cap. But Rover says it has been steadily improving quality

Sound of the fruity VVC was glorious bouncing off Alps

MGF 1.8 VVC

Car run for One year
UK sales to date 11,000
Changes since delivery Improved build quality, wider choice of options and colours
Likes Sweet engine note, spirited performance, lack of scuttle shake, wacky colour
Dislikes Noisy with soft top up, feeble heating system, too many rattles from cabin and hard top

The driving position is not the best, but there is still much fun to be had

since launch, which bodes well for anyone who buys a new one.

Although our original road test generally praised the MGF's handling, I can't get too excited about it. Yes, the car is capable and secure; scuttle shake is virtually non-existent and there's no shortage of grip. But I don't consider it rewarding to drive. It feels a little too heavy and unwieldy for my liking, and the major controls give the impression that they're enveloped in treacle.

The arrival of the Lotus Elise hasn't helped. True, the MGF is more practical and has a far more efficient roof, but the Elise is such good fun every time you get behind the wheel that you can forgive it a lot. The MGF feels rather mundane in comparison.

Having said that, we've had some good times in our MGF. In fact, the last week of its stay with us was spent tearing around Italy and the Alps, mostly with the roof off. The VVC engine sounded better than ever as the noise bounced off the mountains and tunnels, and we continue to be impressed with its spirited performance and quality ride. Some of the more demanding mountain passes revealed a vibration from the front brake discs when they got hot, and the jack proved rather difficult to operate when a wheel change was required, but otherwise the MGF acquitted itself admirably, returning up to 45mpg at times and swallowing a surprising amount of gear.

Some people questioned the need for the optional £1100 hard top, arguing that the soft top was perfectly adequate and that the hard top only restricted our ability to go topless whenever the temptation arose. That's true: if there was any chance of a roof-down opportunity during a journey, the hard top had to stay at home in the shed. But I was thankful for it over the winter months. Not only did it reduce road noise significantly and improve the ambience of the cabin, but its heated glass rear window made all the difference to rear visibility on cold mornings. Apart from its poor fit, its only flaw was that it didn't have rain gutters, letting water fall on the car's occupants when the doors were opened.

Leaving the hard top behind may give you flexibility, but it means you have to accept the limitations of the soft top: extra road noise, poor rear visibility through the plastic screen (on top of an ineffective demister/ventilation system), exposed side hinges on which to bang your head. The MGF is not a great long-distance cruiser in this guise, although our trip to Italy proved that it's not out of the question. At least the soft top is quick and easy to raise or lower.

MGF is practical enough for a European touring holiday for two

My change in attitude towards the MGF is as much of a surprise to me as anyone. I really thought I was going to enjoy every moment of it, but once the summer was over I struggled to maintain my enthusiasm. A roadster that's undemanding and practical may sound like a dream come true, but the MGF fell short in the areas that are most important to me – like driver appeal and driving position – and the list of niggles was too long for comfort. But it's half my fault; if I had more of a passion for open-top cars, things probably would have worked out differently.

Our car was a guest of honour at the MGF's first birthday bash

Handling lacks fun factor; hard top was very welcome in winter

Log book

Name of car MGF 1.8 VVC
Price on arrival £18,795 (£20,560 with options)
Trade value now £17,445
Options Hard top £1095, metallic paint £315, passenger's airbag £355
Costs per mile 24.7 pence (9.5p excluding depreciation)
Average mpg 32.6
Best/worst mpg 47.2/28.3
Servicing 3k miles, free; 12k, £97.11; 24k, £148.25; 36k, £123.57; 48k, £137.72; 60k, £370
Sample labour rates London, £45; Kent, £56; Birmingham, £42; Yorkshire, £38
Running costs Fuel over 20,441 miles, £1570; oil, nil;
Tyres one tyre replaced due to puncture, £89 (205/50 VR15 Goodyear NCT Touring)
Parts costs (including VAT but not fitting) Front bumper, £209.15; headlamp unit, £94; door mirror, £66.97; windscreen, £136.30; wheel rim, £151.57
Insurance Fully comprehensive insurance premium for:
1. 35-year-old married man living in low-risk Swindon, with clean licence and five years' no claims, car garaged: £585.52
2. 25-year-old single man with two speeding convictions and five years' no claims, living in reasonably high-risk Middlesex: £742.56
Faults Rubber seal on hard top replaced under warranty to stop rattling; new pipe fitted between engine air intake and throttle body to cure air flow restriction under duress (no charge)
Other reports 15.5, 12.6, 25.9, 18/25.12 ('96), 5.2, 9.4 ('97)
Where to find out more Rover Group, Warwick Technology Park, Warwick CV34. Tel: 01926 482000

COMPARISON
CATERHAM SEVEN SUPERLIGHT v LOTUS ELISE v MGF 1.8i VVC

THREE-TIER TUMBLE

WHAT MAKES A REAL TWO-SEATER SPORTS CAR? RED RAW MINIMALISM, COMFORTABLE FUN OR SOMETHING SOMEWHERE IN BETWEEN? TONY DRON ON THE CATERHAM SUPERLIGHT, MGF VVC AND THE LOTUS ELISE

COMPARISON CATERHAM SEVEN SUPERLIGHT v LOTUS ELISE v MGF 1.8i VVC

THREE-TIER TUMBLE

Any doubts about the idea of wearing helmets for a road trip in a car were forgotten after the first 20 miles. After all, the Caterham Superlight is hardly any car, the journey from Bucks to Wales no mere scoot round the block. Our transport, for instance, doesn't have a windscreen – not even aeroscreens – just an inch-high smoked Perspex deflector attached to the bodywork just behind the bonnet. Eye protection is as vital as strong neck muscles in this leanest of Super Sevens.

Two things force your head back. First, the startling shove of the 140bhp 1.6-litre K-series engine. Second, the rush of wind at speed. We shouldn't have ditched the head restraints. After over two hours, the added protection from the wind and light rain provided by our helmets is very welcome. Ear plugs, a biker's waterproofs and good gloves aren't luxury items in this uncompromising machine. The Superlight is the most extreme road-legal sports cars I have driven. It sure ain't for shy, retiring types.

We strike out across country, then take the M50. It's brighter now and a bizarre thought flashes across my mind for a split second: "Great graphics", I think to myself, and smile. It's like an inter-active video game, played inside a wind-tunnel. All the elements are there: the glow-

Red raw: highly modified K-series engine hurls the Caterham on to a numbing 7,700rpm cut-out

Ritualistic buckling of the racing harness sets off the adrenaline

76

Gawkworthy Elise is much more than a posing machine. Handling balance and truly potent performance are a joy

Tipping the scales at just over 500kgs, Caterham has a scary power-to-weight ratio. Sixty clocks on in just 4.6sec

Spartan, no-nonsense aluminium-clad cabin is surprisingly spacious. Passenger legroom's prodigious

ing orange Marlboro-McLaren paint job, the matt black headlight cones, front wings bobbing in unison with the wheels, the ribbon of road stretching out ahead, the tiny low-set wheel, the steady blare of the engine and, occasionally in the corner of the 'screen', passenger Henry Bernhard's helmet.

Deeper into Wales we rendezvous with 'the opposition' – two conspicuously modern sports cars, both about the same price as the Caterham but offering strongly contrasting driving experiences. One is an MGF VVC, the other a Lotus Elise. Easy to extreme in three moves.

When I first drove an Elise, I was knocked out by its poise, its wonderful handling, its lightness and efficiency and, frankly, its sheer style. It's one of those 'great from any angle' cars.

But is it an extreme car? By MGF standards, maybe. Then along came the Caterham Super-Seven Superlight. There aren't many cars that make an Elise feel like a comfortable tourer but this is one of them.

True, tall people are accommodated in the Caterham but it's a close fit all around. Then again, everything about the Superlight is close and intimate: it's raucous, it's firm and it's incredibly direct. Twitch and it responds precisely. Instantly. There's no hood and, as we already know, no windscreen. So the answer is, no, the Elise is a well judged compromise. But only the Caterham takes you all the way.

Which begs another question: is the MGF simply too soft and friendly to satisfy? No question it's a great all rounder but there are also times when the MG scores heavily in specific areas.

It may be softer in appearance than the other two but it looks good in this company, hood up or down. Then there's its sophisticated Hydrolastic suspension which is unbelievably good through bumpy, sweeping corners, wet or dry. And the VVC engine is much quicker on acceleration and top speed than you might think, capable of sprinting to 60mph in 8.2sec and 100mph in 22.6.

Yes, the MGF is comfortable, handy for shopping and all that but it is also a proper sports cars

COMPARISON   CATERHAM SEVEN SUPERLIGHT v LOTUS ELISE v MGF 1.8i VVC

MGF has the broadest 2up sports car appeal here. It's much harder to provoke than its rivals but still elicits wide-eyed excitement

THREE-TIER TUMBLE

Cabin shows MGF is clearly the comfortable option here

that will appeal to the majority of enthusiasts in this market. And then, high up a Welsh mountain, when it starts to rain hard the MG driver wins again.

For once I was in the right car at the right time: I just flipped the soft-top over my head, clipped the two toggles in place and pressed the electric window buttons. All done in seconds. I picked up the crossword and filled in a few clues, occasionally glancing up to see Henry fitting the tonneau to the passenger's side of the Caterham before he climbed back into the exposed driving seat and wiped the rain from his wet visor.

Meanwhile, road tester Ollie Peagam struggled with struts and an Allen key to seal the Lotus in the downpour. There's no electric roof on the MGF but the manual set-up is child's play. How soft do you want to be?

The Superlight is altogether less compromising. It will appeal most to a small band of ultra-enthusiasts in search of a kind of ultimate – the odd sprint or hillclimb will almost certainly be on the agenda. The truth is that they will be the kind of people who have a number of cars already and want something exciting for the weekend.

It should satisfy them completely, being so well sorted. The brilliantly modified K-series engine delivers good torque but stills howls past 7,000rpm (7,700rpm is the limit). Caterham's own exquisite short-shifting six-speed gearbox, the excellent Bilstein dampers, phenomenal grip from the Avon FF-racing type tyres (good in the wet too) and powerful, sure-footed braking all contribute fully to the intensely vivid driving experience.

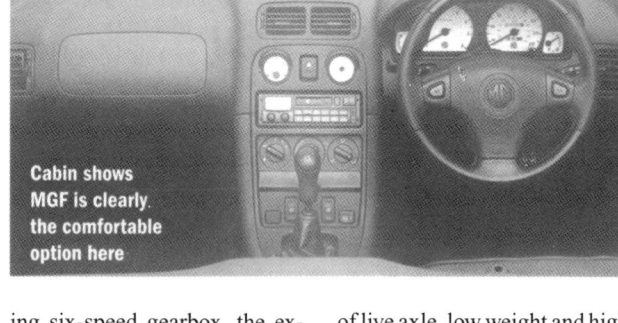

It is blisteringly quick off the mark, catapulting from rest to 60mph in a barely believable 4.6secs. You can throw it on opposite lock but the combination of live axle, low weight and high grip requires concentration and sensitivity to master. I gave myself a sideways surprise the first time I tried that trick but it's fabulous when you get to know it.

Driving the Elise after that, you are aware of the extra space, the longer throw of the five-speed lever and the slightly less direct steering. There's an amazing amount of room inside and it accommodates tall and/or broad people without squeezing them. To drive, it requires a different technique from the Caterham. If you enter a corner too fast it will run wide; clumsily lift off mid-bend and it can snap sideways. But get steering and throttle working harmoniously together and it's a delightfully well-balanced, grippy and safe car.

Its supple, even ride gives the front wheels an easy time and helps make it a supremely effortless consumer of corners. Like all light cars with broad tyres, it's prone to aquaplaning in deep water, but otherwise its behaviour is outstanding: despite that standard MGF engine, low weight and sleek shape give it genuinely potent performance. A 0-60mph time of 6.3sec is a good league division higher than the Rover's 8.2sec.

The MG has to be safe and understandable for every driver and it fulfils that brief remarkably well. Suitably provoked in the wet it will slide its tail but in a progressive, containable way that puts some mid-engined designs to shame. In tight corners, it's more likely to peel away from the apex nose first. That's exactly how it should be.

The Superlight and Elise need to be that bit trickier to provide the deeper level of reward they do. Your mother probably wouldn't like it but it's the sanest choice from this bunch, no doubt about that.

Borders on anodyne from some angles but touches like the bullet-hole filler cap are fun

KEY FACTS — Raw (blue), medium or well done – you choose

Model		Caterham Seven Superlight	Lotus Elise	MG F 1.8i VVC
LIST PRICE	£	17,495	20,950	19,940
Engine				
Cylinder		Inline 4	Inline 4	Inline 4
Capacity	ccm	1,588	1,795	1,796
Power	bhp/rpm	138/7,000	118/5,500	145/7,000
Power to weight ratio	kg/bhp	3.6	6.2	7.8
Torque	lb ft/rpm	115/5,000	122/3,000	128/4,500
Transmission		6sp manual	5sp manual	5sp manual
DIMENSIONS				
Kerb weight	kg	505	733	1,136
Length/width/height	mm	3,100/1,575/800	3,726/1,701/1,201	3,910/1,630/1,270
Wheelbase		2,230	2,300	2,380
Boot volume	litres	120	95	209
Tank volume	litres	36	40	50
SUSPENSION				
Front		Dbl wishbones	Dbl wishbones	Dbl wishbones
Rear		De Dion	Wishbones, trailing arms	Dbl wishbones
Wheels/tyres	front	6Jx13 6x13x21	5.5Jx15 185/55	6Jx15 185/55
Wheels/tyres	rear	6Jx13 6x13x21	7Jx16 205/50	6Jx15 205/50
Brakes	front/rear	Vented/solid discs	Vented discs	Vented/solid discs
PERFORMANCE				
0-40mph	sec	2.3	3.0	3.9
0-50mph		3.8	4.4	5.8
0-60mph		4.6	6.3	8.2
0-100mph		14.5	18.6	22.6
40-60mph (in 4th)		2.9	6.7	9.1
50-75mph (in 5th)		4.7	10.7	14.6
Top speed	mph	129	126	130
BRAKING				
30-0 mph	m	8.9	9.3	9.2
50-0 mph	m	24.4	25.8	25.5
70-0 mph (cold/warm)	m	48.4/_*	50.4/50.2	49.9/50.0
FUEL CONSUMPTION				
Composite	mpg	_**	39.4	40.0
Overall test		30.5	36.2	32.8
EQUIPMENT				
ABS		–	–	£670
Airbag (driver/passenger)		–	–	•/£355
Air condition		–	–	£1,200
Alarm		–	£295	•
Alloy wheels		•	•	•
Immobiliser		•	•	•
Metallic paint		from £764	£690	£315

*Not figured because of wet conditions **fig not published
KEY • Standard equipment – Not available £ Cost option

PERSONAL CHOICE

Irresistible fun, the Caterham Superlight is a brilliant toy nicely set up by a team who know how to take this concept to the very edge. But for every day use? I'm not so sure.

I could live with an MGF. I like the mid-engined feel and the impressive high-speed ride and handling. It's easily the best MG of modern times and close to being the optimum sportscar for the masses.

But I would take the Elise. I love the concept, the engineering, the look. And a big plus is that, even at 6ft 5ins, I don't feel the wrong size for it. It's exciting but not outrageously extreme – the ideal compromise, I reckon. This time, Lotus has got it right. Exactly right.

Tony Dron

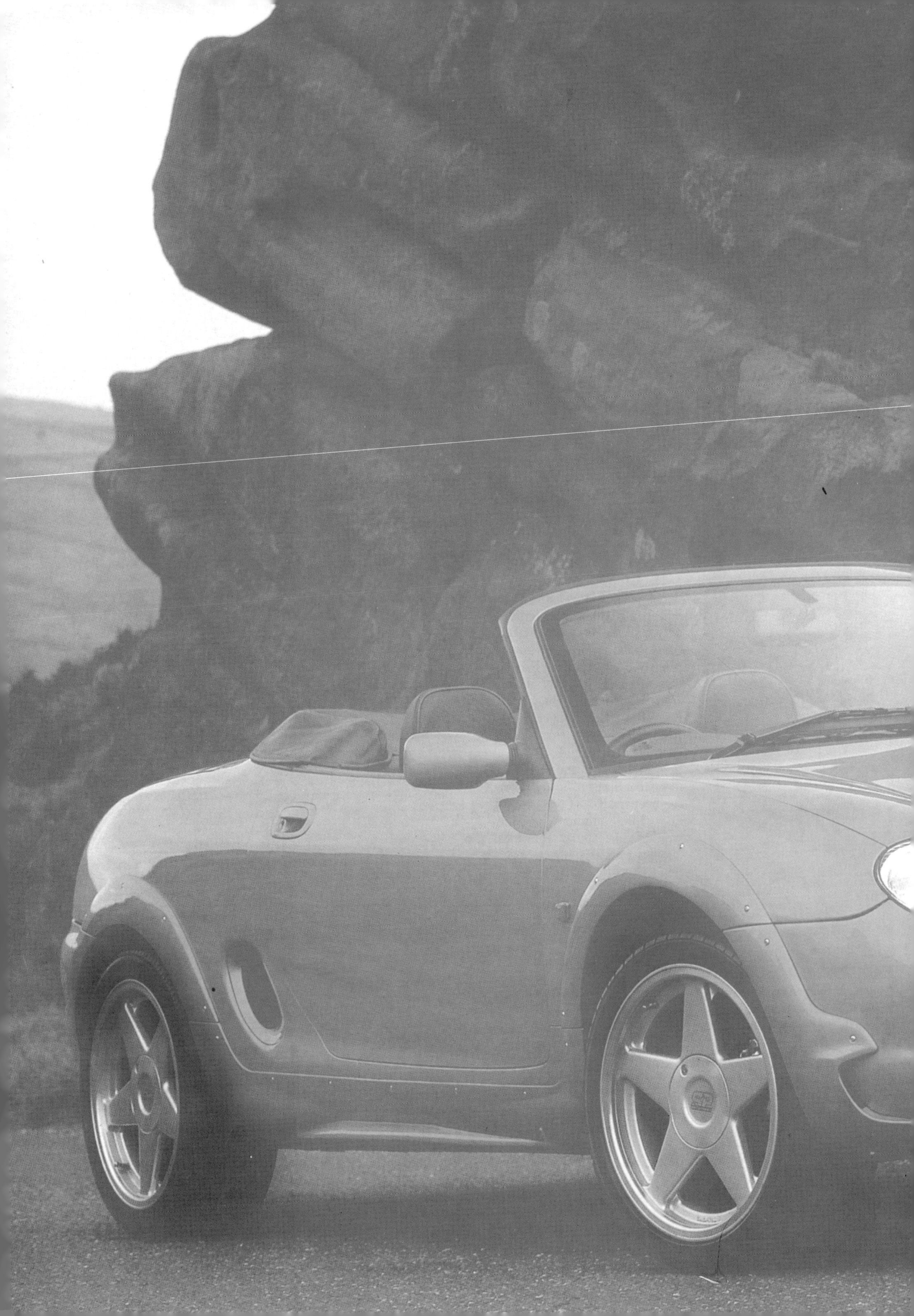

MG FF

FIRST DRIVE | MGF CHEETAH The world's first supercharged MGF hasn't been built by the factory, but by an enterprising Rover dealership. Andrew Golby reports

This is the story of a Rover dealer who has beaten the factory to building the world's first supercharged MGF. This week, Rover revealed a supercharged MGF concept at the Geneva show; but Nottinghamshire Rover dealer Stephen Palmer Ltd already has its car on sale.

This isn't the first time that a pioneering privateer has got the jump on MG. Ken Costello did much the same with the MGB, fitting it with a V8 in 1970, three years before the factory's car. Palmer won't enjoy quite such a long honeymoon; we expect a Rover-built blown MGF to be on the roads before the year's end. But the £29,500 Cheetah will forever be the first.

It has been conceived from scratch in just seven months by SP Performance, Palmer's tuning arm. SP was set up just three years ago (although Palmer has been a dealer since 1988), offering exhaust kits for standard 1.8-litre MGFs. At the time, demand for VVC models hugely outstripped supply. Palmer's performance modification was a hit, and the company has since become the country's top MGF dealer, selling 250 new cars a year. Having spoken to many customers wanting something faster, but no less refined, Palmer took the plunge – and the Cheetah is the result.

It isn't merely supercharged; the Cheetah has uprated brakes, revised suspension, an all-new exhaust system, bespoke wheels and a host of styling changes including new front and rear bumpers and wheelarch extensions. Unique dials are fitted inside, with a pair of backlit eyes on the instrument panel. A steering column switch allows them to be turned on and off as you please. For those who don't want the full Cheetah treatment, its components will be available in sub-kits. The supercharger should cost about £4000, the exhaust system £300, the body kit £2500 and the brakes £400. This is an awful lot more than a dressed up MGF.

But why forced induction? The 200bhp produced by this

Cheetah developed in seven months

blown 1796cc VVC engine could almost certainly have been achieved by a normally aspirated unit. Indeed, 190bhp has already been coaxed from the K-series, but anyone who's been near a Caterham Superlight R knows that the engine is hardly a model of refinement. And for this reason, a supercharger seemed to be the best option. The plan was not to build a super-quick, no-compromise MGF, but to retain the car's character in a faster package.

Turbo Technics is responsible for supercharging the K-series and remapping the electronics. The new Danish-built Rotrex supercharger is state of the art. Not only is it compact and lightweight (important considering the MGF's cramped engine bay), it uses turbo technology to improve efficiency and power. It is a centrifugal blower and, uniquely, uses epicyclic friction gear to allow the vanes to spin at up to 100,000rpm. Airflow has been further improved by piping cold air from each side intake directly into the engine bay.

Power may eventually rise to 210bhp

State-of-the-art supercharger fitted to VVC K-series by Turbo Technics

at 6800rpm. The car we drove was a prototype, not quite running under full pressure. Meanwhile, torque stands at 162lb ft; better still, it remains at that figure all the way from 3000-7500rpm. And if you like the whistle of a supercharger, you'll love the noise this car makes. The new exhaust system contributes 12 per cent to the overall power increase and sounds much fruitier than the standard item. But true to SP's aims, it isn't *too* noisy.

It's faster, of course. Palmer quotes a sub-6.0sec 0-60mph time, a claim we can't quite back up from our drive. The extra pace is deceptive; the car remains very easy to drive in town and has the same road manners as other MGFs. The engine never sounds as if it's working as hard as it really is, but corners arrive noticeably quicker than in a standard VVC. Throttle response is usefully urgent, although the revs are

Alloys are unique to MGF Cheetah

Only blown MG on sale... for now

Prototype down on power, but Cheetah should offer at least 200bhp. Doesn't feel as agile as standard MGF, but level of grip is much higher

hope those that have will perform much better.

The MGF's chassis in standard tune is one that we have always admired for its versatility, even if we've wondered how it would feel if it was a little stiffer. SP Performance has avoided the temptation. Instead, it has lowered the car slightly, changed the position of the bumpstops and actually softened the Hydragas springs. Its reasoning becomes clear when you realise that the Cheetah runs on big 17in wheels with low-profile Yokohamas – 215/40s on the front and 215/45s at the rear, compared with 185/55s and 205/50s as standard.

Differences are noticeable straight off. Unfortunately, the Cheetah loses some agility; its steering is noticeably heavier and feels a little clumsy compared with the standard car. In its favour, the Cheetah's grip levels are well up on other

Exhaust boosts power, sounds fruity

Intakes pipe air straight to engine

MGFs and the chassis is every bit as secure. The ride quality is little worse; only large potholes send a pronounced thud through the cabin.

While the Cheetah obviously needs fine tuning, it will appeal to those looking for something a little different. Palmer is to be admired for attempting such a bold venture and sinking some £100,000 into the project. He expects to sell 120 Cheetahs every year. And as the dealer who sells more MGFs than anybody else, he has a better chance of pulling this off than anybody.

curiously slow to drop once you are off the accelerator. The car we drove had an unusually sticky gearchange as well; MGF gearboxes are hardly renowned for their slickness, but this one felt particularly notchy. No changes have been made to the gearbox, so perhaps the problem was specific to this car.

To cope with the extra performance, Palmer has fitted Mintex competition discs and pads. Those on our test car were not at their best and stability under repeated hard braking was poor. Apparently, the uprated parts had completed around 1500 miles, but had not been properly bedded in. We

Supercharged MGF's dials feature backlit pair of trademark Cheetah eyes

FACTFILE

MGF CHEETAH

HOW MUCH?	
Price	£29,500
On sale in UK	March

HOW FAST?	
0-60mph	5.9sec
Top speed	150mph

HOW THIRSTY?	
Urban	25mpg
Extra urban	45mpg
Combined	33mpg

HOW BIG?	
Length	3914mm
Width	1780mm
Height	1260mm
Wheelbase	2375mm
Weight	1235kg
Fuel tank	50 litres

ENGINE
Layout 4 cyls in line, 1796cc
Max power 200bhp at 6800rpm
Max torque 162lb ft at 3000-7000rpm
Specific output 111bhp/litre
Power to weight 162bhp/tonne
Installation Mid, transverse, rear-wheel drive
Made of Alloy head and block
Bore/stroke 80/89mm
Compression ratio 10.5:1
Valve gear 4 per cyl, dohc
Ignition and fuel Remapped MEMS electronic ignition, multi-point fuel injection, Rotrex supercharger

GEARBOX
Type 5-speed manual
Ratios/mph per 1000rpm
1st 3.17/5.9 **2nd** 1.84/10.1
3rd 1.31/14.2 **4th** 1.03/18.1
5th 0.77/24.1 **Final drive** 3.94:1

SUSPENSION
Front Double wishbones, Hydragas springs, anti-roll bar
Rear Double wishbones, Hydragas springs, anti-roll bar

STEERING
Type Rack and pinion, power assisted
Lock to lock 3.1 turns

BRAKES
Front 240mm vented discs
Rear 240mm solid discs
Anti-lock Standard

WHEELS AND TYRES
Size 7.5Jx17in **Made of** Alloy
Tyres 215/40 ZR17 (f), 215/45 ZR17 (r) Yokohama A520

All manufacturer's claims

VERDICT
Admirable effort, but too many flaws to be considered completely successful.

WIND CHEE

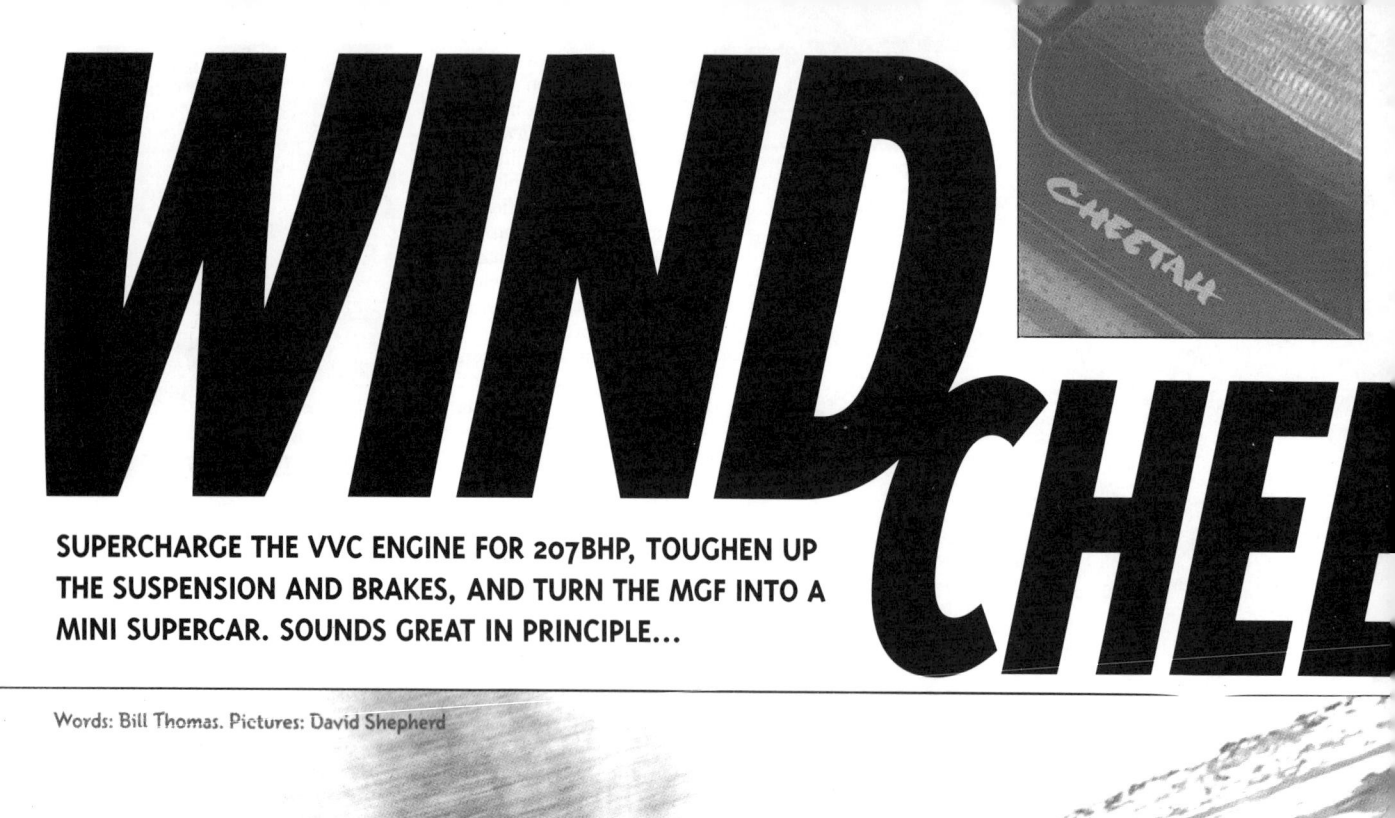

CHEETAH

SUPERCHARGE THE VVC ENGINE FOR 207BHP, TOUGHEN UP THE SUSPENSION AND BRAKES, AND TURN THE MGF INTO A MINI SUPERCAR. SOUNDS GREAT IN PRINCIPLE...

Words: Bill Thomas. Pictures: David Shepherd

Sorry, SP Performance, but your supercharged MGF Cheetah just didn't quite agree with me. It was an occasion when I really wanted to fall in love with the car, to cherish it and enjoy its company, to remember it forever and recall the fun times I had with it, the love and romance and throbbing passion which had etched itself onto my memory. As it was, I'll forget it over the weekend.

The reason I'm describing a car in such sick-making terms is probably the same reason the Cheetah was in trouble before I'd set eyes on it. John Barker had kindly lent me 'his' long-term test Impreza Turbo to bomb around in for a week. Note the inverted commas around 'his', because it's not really 'his'. It's 'mine'. It told me that it loved me more than it loves him, that it wants to be with me for eternity. Really.

Much has been written about the Scooby in the hallowed pages of this magazine, so I won't bore you with anything more, except to say that it is the best car I have ever driven. It's so good it makes me walk differently. I adopt a sort of bowed stance, arms and upper body unmoving, shaking my head in wonder, looking at the floor and shuffling forward slowly. That's if I'm not walking backwards, admiring the thing, listening to it tick as it cools. Arghh!

So this Cheetah, at a cool £30,000, ten grand more than the Scooby, will need to be pretty impressive. And, on paper, it seems to have much to recommend it: it has the potential to be a fantastic little car, a mini-Ferrari almost. The engine is a supercharged version of the already-excellent 1.8 VVC, fettled by forced-induction experts Turbo Technics, which ups the power from the standard 143 to a claimed 207bhp, with peak torque boosted from 128 to 162lb ft. These figures put the Cheetah squarely into Impreza territory. In a rear-wheel-drive, mid-engined car weighing only 2337lb (400lb less than the Scoob), it should be quick and interesting, especially in the wet...

The suspension is also heavily revised, benefiting from experience gained with the Hydragas competition set-up used in French and Japanese racing series. The brakes are fully uprated with racing pads and discs developed by Mintex, linked with the MGF's Bosch ABS system.

In fact, SP has done a superb job all round: witness the body styling. Developed in conjunction with Krafthaus of Nuneaton (which has worked on projects for Jaguar and Aston Martin), the Cheetah's new front

WIND CHEETAH

'We speared through the freezing Peak District with the top down'

and rear bumper units give the MG the aggressive look it always deserved: the front end features a deeper main air intake incorporating two driving lights, an adjustable splitter to help reduce nose lift, and two extra air ducts in the lower bib flanking a pair of smaller spotlights; at the rear, extra air vents and a re-profiled lower bumper also toughen things up.

Bulging wheel arches and sculpted side skirts help the car's stance, though I wasn't sure about the exposed bolts on its flanks (these are optional). Sitting on superb Azev 17in five-spoke alloys and Yokohama 215/40 tyres, the Cheetah is a fine aesthetic development of the standard car – aggressive, purposeful but with a subtle quality that won't be outdated in five minutes. Pedestrians pointed. Other MG drivers were impressed. That's all you can ask – full marks, SP.

But the proof, as always, is in the driving. And I couldn't. My 6ft 2in frame simply wouldn't fit. Tall drivers beware! The steering wheel was mounted so low it was in my groin, and moving the seat all the way back on its runners didn't help. My hands were still smacking my knees on anything more than a quarter of a turn. SP has moved the column upwards, but it clearly wasn't enough. It's looking at lowering the seat height as a final solution.

On the move, the polite warning from Stephen Palmer, boss man of SP, was ringing in my ears: it's wet, so be careful. And I was… In my mind, the supercharger would give me a surge of extra torque low-down, a lag-free dollop of spunk that would have the rear wheels spinning the instant I nailed the throttle, then continue to give the engine extra thrust all the way to the 7500rpm redline.

Funny, then, that it didn't happen that way. For some reason, the engine seemed reluctant to build revs rapidly. I was told that it was quite new and tight, but even taking that into account, I expected to be more impressed by the acceleration, especially between 3000 and 7000rpm. The previous press demonstrator had been clocked by another magazine at 5.5secs to 60mph and 12.9 to 100mph – impressive indeed. We didn't have a chance to figure the car ourselves, but subjectively it didn't feel like this example would get close to those times. There wasn't a lot of urge down low, and this was combined by a further reluctance to rev higher up the range. Photographer Shepherd and I were puzzled – at full throttle in second gear, we waited for the needle to ram itself toward the redline and the seats to push through our backs, but it never happened. Yes, it feels quite fast, but it was as if the boost had been turned down: in essence, if this engine does develop over 200bhp, then I don't like the way it delivers it.

Perhaps the engine wasn't quite on song, perhaps it was too new, maybe the gearing was upset by the larger diameter wheels, maybe I should have driven the other demonstrator. Whatever, it left me wanting

SP allows you to take your pick from its performance and styling upgrades (we'll take the wheels and the exhaust, thank you). Alternatively you can go the whole hog (or rather the whole Cheetah) and have the lot

to throw it onto a dyno to confirm its power, to run it against another car with a similar power-to-weight ratio. What it didn't leave me was scared.

Everything else about the car seems fine: the gearchange is quick and easy, the brakes have tremendous power and a progressive pedal feel, and the handling is sharp and entertaining. The Cheetah turns in crisply with very little roll, and holds on forever thanks to its huge Yokohama pawprint. When traction does break, understeer is your first warning, and lifting off mid-bend has no ill effects, even in the wet, though I was being circumspect with my speed. Best of all, the ride quality hasn't suffered a great deal from the firmer set-up – it's chunky, but never jarring.

So, the chassis is a good thing: all the more reason, then, to reap the benefits of a meaty and willing little engine behind your back. Pity, eh? And it was: we speared through the freezing Peak District with the top down and the exhaust note became a thing of utter joy. This is one of the finest rasps I've heard from any four-cylinder engine, a beautiful high-pitched scream, and we found ourselves revving the thing deliberately in small villages just for the heck of it. That SP sports exhaust system should be fitted to every MG on earth.

But for 30 big ones, the Cheetah doesn't quite justify itself as an overall package – there are plenty of motors we'd consider first before spending that sort of money, the Impreza clearly being one of them, the £22,369 Fiat Coupe 20V Turbo another.

For a touch of MG individuality, SP can supply various combinations of upgrades to suit your needs and budget, but it's keen to promote the Cheetah as a model in its own right. From this perspective, the car I drove is disappointing, though not through any fault of the handling, braking or styling; it's just that the engine in the test car was too weak.

After returning Mr Cheetah to its owners, I hopped into my Impreza and was soon reminded of exactly what 200+bhp should feel like. I took the long way home, grinning like a loon.

Thanks to Stephen Palmer Limited of Long Eaton (0115 972 2321) for lending us the car and rescuing us from the Derbyshire Peaks after snapper Shepherd locked the keys in the boot and we nearly died of hypothermia. It's tough on the edge.

SPECIFICATION

Engine: Rover 1.8i VVC with Rotrex supercharger running .45 bar boost. Remapped engine management system. SP Performance stainless steel exhaust system

Max power: 207bhp at 5500rpm

Max torque: 162lb ft at 4500rpm

Top speed: 150mph (electronically limited) 0-60mph: 5.5secs

Bodywork: Re-styled front and rear bumpers, side skirts and wheel arches. Alloy fuel filler cap

Suspension: Double wishbones; uprated race-developed Hydragas springs

Brakes: Mintex anti-fade pads and competition discs

Wheels and Tyres: Azev 17in five-spoke alloys. Yokohama 215/40s front, 215/45s rear

Interior: SP Performance 'Cheetah' instrumentation: Custom gear knob, leather trim

Insurance group: 14

Basic price: £29,500 otr

BUYING USED

MGF
[1995 ONWARDS]

Thirty years after British Leyland cancelled the mid-engined MGB replacement, Rover sprang the mid-engined MGF on the world and re-established MG as a major sports car maker. Not only does the model follow the MG spirit of tweaked production components to create a sporting road car capable of strong track performances, but it carries enough traditional MG cues to lure back long-lost previous owners. It boasts a thoroughly modern package which also has great appeal to a new generation of potential owners. Well made, cheap to run, involving to drive and oozing charisma, if you drive one, you'll want one. **Richard Wilsher**

RICHARD PARSONS

Check nose for excess stone chips

Coolant requires expert refilling

VITAL STATISTICS

PRICE RANGE £15,500 (95N 1.8i) to £21,000 (98R VVC)
LAUNCH DATE Sept '95. 1.8i has 120bhp K-series engine, airbag, alloys, electric windows, alarm/immobiliser. Options include dual airbags, hard-top and electric power steering. 1.8i VVC has 143bhp K-series engine. Specification as 1.8i but with half leather
PROS Prices becoming more sensible, cars reliable and well made
CONS Dealers scarce, abused cars in private ads for dealer money

CHECKING IT OUT

Telling 120bhp 1.8 and 143bhp VVC variants apart is not easy. As a guide, the VVC has five-spoke alloys versus the 1.8's chunky six-spoke items, front fog lights and a third LED brake light mounted under the boot lip.

The previous owner may also have plumped for the £1450 colour-keyed hard-top. There are four slightly different hard-tops. Make sure the one you're looking at was supplied new with the car, to avoid possible leaks and paint damage. Remove it and check the condition of the hood fabric, mechanism and stitching.

Uncaring owners are likely to be the MGF's biggest problem. Buy a scruffy purple '95 N-plater from the private classifieds with a twee registration number and no service history and you're cruising for a financial bruising. Sports cars are meant to be driven with verve but the MGF's long service intervals and thinly spread dealers, coupled with constant driver abuse from the sort of people who ignore any signs of protest, are probably the main reasons for tales of MGF unreliability.

Spotting a banged-about

Check £380 rear silencer is okay

BRITISH SPORTS CAR SPECIAL

Panel shut lines should be perfect

MGF is not that difficult. Examine the car carefully for true shut lines, even panel fit, damaged wheels, and driver's seat and door aperture damage. Expect minor stone chips, especially on green cars' soft pigments, but no rust. Ride height is important for tyre ▶

Trim is well screwed down so suspect over-rattly cabin; roof leaks not unknown, but wet floor unacceptable

89

BUYING USED

Flexible exhaust joint can be source of annoying rattle from under the car

Water coolant pipes under car poorly protected, vulnerable to speed humps

wear. The MGF's hydrolastic suspension is easily tampered with, so avoid a car that rides suspiciously low.

Few cars will have made the 36,000-mile major service yet, so check records very carefully. The MGF is not difficult for an independent garage to service properly so there is no excuse for missed services. Instruments and electrics come straight out of the Rover 200 and are well proven. Check the alarm/immobiliser system works; the plipper has a short range.

The MGF is essentially an open car and although most squeaks and rattles can be fixed with careful trim adjustments, some will always be there. Drips of water through the hood, especially if the car is parked on a steep camber, are likely with all convertibles, so don't get too upset. Properly serviced early versions should have had their windscreen and boot seals updated by the factory.

Check the brake discs for wear. They are visible through the wheels. Discs and pads are expensive at £200-£250 all round. Now get under the car. Some engines have been known to leak oil from the clutch housing. It is not cheap to fix. Early examples of the complex cooling system joint at the front of the engine leaked but have since been modified. The water pipes from engine to front radiator are vulnerable to damage.

An annoying rattle under the car is usually the exhaust front flexible joint and early cars may be due for a £380 rear silencer. Clutches and gear linkages may be troublesome, so pay attention to what should be a neat, slick change with no graunching.

Underside inspection vital but not too many worries ★★★

PARTS & SERVICING

The MGF's service intervals are every 12,000 miles so keeping one in top condition isn't going to break the bank. MG dealer SMC Rover (01895 232425) charges £130 for the 12,000-mile/one-year service, £150 for the 24,000/two-year and £183 for the major 36,000/three-year.

However, as we stress with all long service intervals, do check if your car begins to act differently. Most likely problems will be battery failure, so sluggish starting or a flashing battery light should be checked. Routine servicing and maintenance are easy but electronic gremlins will need Rover's diagnostic equipment.

SERVICE COSTS

Parts prices include VAT, but not fitting

● Front discs	£71
● Front pads	£63
● Rear discs	£82
● Rear pads	£88
● Headlight	£112
● Headlight glass	£51
● Rear light	£62
● Windscreen	£145
● Door mirror	£75
● Rear silencer	£380
● Front shock	£57
● Fuel filter	£14
● Hard-top	£1450

Sensible maintenance costs but rear silencer a touch steep ★★★

SNAP JUDGEMENT

Keenly priced at £16,750 with hard-top and VVC wheels, this 1.8 96P sales car from Wilsher's Garage had done 12,000 miles and came with a six-month/6000-mile full warranty. Everything but the glass and radio were covered. Minor stone chipping could be found on the front, but otherwise the car looked new. Tootling about the Cambridge countryside with a pleasant snorting from the stainless exhaust at 25mph in fifth (the engine is very flexible) revealed the car to have been carefully prepared. There were few squeaks and rattles, while standing on the loud pedal produced a satisfying roar and kick in the back as the car burst into life. Wonderful brakes were the icing on the cake.

Shows a used MGF can feel like new so buy nothing less ★★★★

BUYING USED

Check hood fabric for tears and deep creases; to check its water-tightness run vehicle through a car wash prior to agreeing the deal – if seller will let you

WHAT THEY COST

Although there are now plenty of used MGFs around, the small difference in price between an MG approved used example or one from an independent specialist, and one bought in a private sale where no warranty is offered means you have to seriously consider whether saving a few quid buying privately is worth the risk.

Similarly, you will find import agents sourcing cars from the continent advertising new MGFs at prices equivalent to used UK cars, but is it really worth all that hassle when MG dealers such as SMC Uxbridge have about 25 MG approved cars of all specifications and prices just waiting to be driven away? Every one of its approved used MGFs carries a one-year parts and labour warranty with AA cover, Homestart and, if you're not happy with the car, the chance to exchange it within 30 days. Cars are also checked with HPI for finance and accident history, given a 120-point mechanical inspection, and serviced and updated with the latest product improvement components. On an early car that last benefit could be worth as much as £900.

The MGF is not especially colour sensitive, but green seems to chip more easily than most while purple is frowned on by sellers. Independent MGF dealer John Wilsher, who helped us with our research, says it's the only colour that struggles to sell and he wouldn't stock another.

Don't automatically assume private cars are the best value ★★★

PRICES WE HAVE SEEN

YEAR	ENGINE	MILES	SPEC	PRICE	SOURCE
●95N	1.8	16k	-	£15,950	Trade
●95N	1.8	7k	Hard-top	£16,995	MG app
●96N	1.8	7.5k	CD	£16,500	Private
●96N	1.8	7.5k	Hard-top	£16,995	MG app
●97P	1.8	13k	-	£16,295	MG app
●95N	VVC	30k	-	£17,000	Private
●96N	VVC	28k	-	£17,995	MG app
●96P	VVC	10k	-	£19,295	MG app
●97R	VVC	9k	Air con	£21,495	MG app
●97R	VVC	5k	-	£19,500	Private

LIVING WITH ONE

Autocar managing editor Allan Muir ran an MGF VVC for 12 months and 20,000 miles. The closest it came to giving trouble was during a performance test when the engine ran out of puff at 4000rpm. Muir traced the cause to a faulty air hose to the throttle body. Rover fitted a new one and cured the problem.

The only thing to fall off was the passenger's door mirror adjustment lever. The hard-top rattled from new but the soft-top never leaked, even in the heaviest downpours.

Would he have another MGF? No. Nothing wrong with the car – although he says it lacked agility and understeered too eagerly – more his dislike of soft-tops in our fickle climate. "You really have to want a soft-top to enjoy one," he says.

Reliable, water-tight but be sure you really want one ★★★★

INSURING IT

Gone are the days when sports car insurance cost as much as the car. Effective immobilisers mean that city dwellers once at risk from thieves will find their MGF cheaper to insure than those in the country tempted to drive too hard and too fast.

It's a sports car so expect no favours unless you live in town ★★★

INSURANCE COSTS

1.8i		
Age	Location	Fee
45	Matlock	£226
25	Matlock	£387
45	Guildford	£241
25	Guildford	£345
VVC		
45	Matlock	£289
25	Matlock	£530
45	Guildford	£303
25	Guildford	£474

SPECIAL THANKS TO: Wilsher's Garage Ltd, Wimpole, Cambs (01223 207226); SMC Rover, Uxbridge (01895 232425); Wimpole Hall, Cambs (01223 207257)

Other cars in the series: Peugeot 205 GTi 4.9.96; VW Golf GTi Mk II 16.10.96; BMW 3-series 8.1.97; Range Rover 25.6.97; Jaguar XJ6 (1986-94) 3.9.97; Mercedes-Benz 190 12.11.97; VW Corrado 7.1.98; Fiat Cinquecento 4.2.98; Rover 200 18.3.98

OFFICIAL TECHNICAL BOOKS

Brooklands Technical Books has been formed to supply owners, restorers and professional repairers with official factory literature.

Workshop Manuals

Model	Original Part No.
Midget TC (instruction manual)	
Midget TD / TF	AKD580A
MG M to TF 1500 (Blower)	XO17
MGA 1500, 1600 & Mk. 2 (SC)	AKD600D
MGA 1500, 1600 & Mk. 2 (HC)	AKD600D
MGA Twin Cam	AKD926B
Midget Mk. 1, 2 & 3 & Sprite	AKD4021
Midget 1500	AKM4071B
MGB & MGB GT	AKD3259 & AKD4957
MGB GT V8 Supp.	AKD8468
MGC	AKD7133/2

Parts Catalogues

Model	Original Part No.
MGA 1500 (HC)	AKD1055
MGA 1500 (SC)	AKD1055
MGA 1600 & Mk. 2 (HC)	AKD1215
MGA 1600 & Mk. 2 (SC)	AKD1215
Midget Mk. 2 & 3	AKM0036
MGB Tourer, GT & V8 (to Sept. '76)	AKM0039
MGB Tourer & GT (Sept. '76 on)	AKM0037

Owners Handbooks

Model	Original Part No.
Midget TD	
MG Midget TF & TF 1500	AKM658A
MGA 1500	AKD598G
MGA 1600	AKD1172
MGA 1600 Mk. 2	AKD195A
MGA Twin Cam	AKD879
MGA Twin Cam (3rd edn.)	AKD879B
Midget Mk. 3 (pub. '73)	AKD7596
Midget Mk. 3 (pub. '78)	AKM3229
Midget Mk. 3 (US) (pub. '71)	AKD7883
Midget Mk. 3 (US) (pub. '76)	AKM3436
Midget Mk. 3 (US) (pub. '79)	AKM4386
MGB Tourer (pub. '65)	AKD3900C
MGB Tourer & GT (pub. '69)	AKD3900J
MGB Tourer & GT (pub. '74)	AKD7598
MGB Tourer & GT (pub. '76)	AKM3661
MGB GT V8	AKD8423
MGB Tourer & GT (US) (pub. '68)	AKD7059
MGB Tourer & GT (US) (pub. '71)	AKD7881
MGB Tourer & GT (US) (pub. '73)	AKD8155
MGB Tourer (US) (pub. '75)	AKD3286
MGB (US) (pub. '79)	AKM8098
MGB Tourer & GT Tuning	CAKD4034L
MGB Tuning (1800cc)	AKD4034
MGC	AKD4887B

ALSO AVAILABLE: 180 page 'Glovebox' size owners' workshop manuals:
MGA & MGB & GT 1955-68
MG Sprite & Midget 1, 2, 3, 1500 1958-80
MGB & GT 1968-81

Note: SC - Soft Cover   HC - Hard Cover

From MG specialists or, in case of difficulty, direct from the distributors:

Brooklands Books Ltd., PO Box 146, Cobham, Surrey KT11 1LG, England
Phone: 01932 865051   Fax: 01932 868803
Brooklands Books Ltd., 1/81 Darley Street, PO Box 199, Mona Vale, NSW 2103, Australia
Phone: 2 9997 8428   Fax: 2 9979 5799
Car Tech, 11481 Kost Dam Road, North Branch, MN 55056 USA
Phone: 800 551 4754 & 651 583 3471   Fax: 651 583 2023